Varieties of Atheism in Science

Varieties of Atheism in Science

ELAINE HOWARD ECKLUND AND
DAVID R. JOHNSON

OXFORD
UNIVERSITY PRESS

OXFORD
UNIVERSITY PRESS

Oxford University Press is a department of the University of Oxford. It furthers the University's objective of excellence in research, scholarship, and education by publishing worldwide. Oxford is a registered trade mark of Oxford University Press in the UK and certain other countries.

Published in the United States of America by Oxford University Press
198 Madison Avenue, New York, NY 10016, United States of America.

Library of Congress Cataloging-in-Publication Data
Names: Ecklund, Elaine Howard, author. | Johnson, David R., 1977– author.
Title: Varieties of atheism in science / [Elaine] Howard Ecklund
and David R. Johnson.
Description: New York, NY : Oxford University Press, [2021] |
Includes bibliographical references and index.
Identifiers: LCCN 2021001995 (print) | LCCN 2021001996 (ebook) |
ISBN 9780197539163 (hardback) | ISBN 9780197539187 (epub)
Subjects: LCSH: Atheism. | Atheists. | Scientists—Religious life. |
Religion and science.
Classification: LCC BL2747.3.E25 2021 (print) | LCC BL2747.3 (ebook) |
DDC 211/.80885—dc23
LC record available at https://lccn.loc.gov/2021001995
LC ebook record available at https://lccn.loc.gov/2021001996

DOI: 10.1093/oso/9780197539163.001.0001

1 3 5 7 9 8 6 4 2

Printed by Sheridan Books, Inc., United States of America

Contents

Acknowledgments

This book was a labor of love and doubt. In 2011, Elaine sat with an atheist biologist in the United Kingdom as she began pilot interviews for what would ultimately become the Religion Among Scientists in International Context study (RASIC)—the most comprehensive international study of scientists' attitudes toward religion ever undertaken. By the time our team finished the first book based on the RASIC data in 2019, *Secularity and Science: What Scientists Around the World Really Think About Religion,* we had done over 22,000 surveys and over 600 in-depth interviews; we were satisfied and we were tired. But over the years our own minds had changed about who atheist scientists were; we came to doubt our own assumptions and we recognized that an important story about the varieties of atheism in science needed to be told. In the pages that follow, we endeavor to accomplish this goal, using richly detailed data to explain the pathways that led scientists to atheism, their diverse identities as atheists, and their views of science, meaning, and morality.

The book would not have been possible without a community of scholars with whom we have been fortunate to work. We are grateful for the undergraduate students, graduate students, post-baccalaureate fellows, postdoctoral fellows, research staff, subcontractors, and others who contributed to this book and supported its completion in a variety of ways. In particular, we would like to thank Colton Cox, Rose Kantorczyk, and Michael McDowell; Dan Bolger, Di Di, Simranjit Khalsa, Sharan Mehta, Esmeralda Sánchez Salazar, and Sandte Stanley. Alex Nuyda provided design assistance for the cover. Work like this would not be possible without Laura Achenbaum and Hayley Hemstreet—and much less fun. Special thanks to Laura Johnson for her statistical work and to Bethany Boucher and Heather Wax for their assistance with manuscript preparation and editing. Thank you to Cynthia Read, our editor, for her endless support throughout this process and for being a champion of this work.

We are grateful for generous funding from the Templeton World Charity Foundation, whose support of the original research made this book possible (Grant #TWCF0033/AB14). Thank you also to the Templeton Religion Trust (Grant #TRT0157 and Grant #TRT 0203) for funding related to analysis and writing. Special thanks to Christopher Brewer, Kara Ingraham, and Christopher Stewart.

This book was written during an eventful period in all of our lives, often meaning work and family were more intermingled than ever. We are deeply thankful to our families for allowing us the space to carry out this work and enriching our lives in the process.

1

Why Study Atheism among Scientists?

Americans generally view atheists as immoral elitists, aloof and unconcerned with the common good. Americans are reluctant to vote for atheists and do not want their children to marry atheists.[1] Atheists have been, in the words of one historian, "a much-maligned minority" throughout American history.[2] Yet, in the U.S. as well as the U.K., some atheists have recently become celebrities.[3] Over the past 15 years, this group—neuroscientist Sam Harris, philosopher Daniel Dennett, the late journalist and philosopher Christopher Hitchens, and evolutionary biologist Richard Dawkins, sometimes referred to as "the four horsemen" of atheism—has appeared widely in the media and public debates, gained a large online following, and written a number of bestselling books arguing that there is an indelible conflict between religion and science. Dawkins, Oxford University's first Simonyi Professor of the Public Understanding of Science, is the most famous atheist of all; in one survey Elaine conducted with Christopher P. Scheitle,[4] they found that more than 20 percent of Americans had heard of him. His 2006 book *The God Delusion* stayed on *The New York Times'* list of top 10 bestselling titles for 16 weeks and has sold millions of copies. Together, the writing, speeches, and public reach of Dawkins and these other "New Atheists" launched a particular version of atheism into the mainstream public discourse to popular acclaim.[5]

Stories from our own lives provide insight as to what that particular version of atheism is. During her research on what religious people think about scientists, Elaine attended a Bible study

Varieties of Atheism in Science. Elaine Howard Ecklund and David R. Johnson, Oxford University Press.
© Oxford University Press 2021. DOI: 10.1093/oso/9780197539163.003.0001

in a rural church in upstate New York. When she told a woman there that she was a doctoral student at Cornell University, the woman responded, "yuck. I wouldn't want my children to attend Cornell." She told Elaine she worried that if they went to Cornell, her children might be exposed to scientists who would take them away from their faith. David remembers some friends who grew up in evangelical Christian families that practiced daily morning devotionals together. One recently "came out" as an atheist; his wife ultimately followed later (after many uncomfortable discussions with her spouse), and the issue has been extremely divisive within their broader families, who attend the same church.

New Atheism has some elements that are genuinely different from earlier forms of atheism, according to Steven Kettell,[6] a professor at the University of Warwick. New Atheism, he explains, is a predominantly "Anglo-American phenomenon" centered on the works of the aforementioned "four horsemen." He points out several common threads New Atheists have in their writing: 1) they place emphasis on science as the only or the superior way of knowing, 2) religion is judged based on scientific evidence and found wanting, and 3) religion is not simply wrong but is pathological and even dangerous. In other words, New Atheism is deeply connected with science and hostility toward religion.

"Faith is the great cop-out, the great excuse to evade the need to think and evaluate evidence. Faith is the belief in spite of, even perhaps because of, the lack of evidence,"[7] Dawkins has said. In his book *The Selfish Gene*, he writes that, "Religion is capable of driving people to such dangerous folly that faith seems to me to qualify as a kind of mental illness."[8] Dennett has referred to religious people as being "disabled" by their adherence to faith,[9] writing specifically: "Suppose you believe that stem-cell research is wrong because God has told you so. Even if you are right—that is, even if God does exist and has, personally, told you that stem-cell research is wrong—you cannot reasonably expect others who do not share your faith or experience to accept that as a reason. The fact that your faith is so strong that you cannot do otherwise just shows (if you really can't) that you are disabled for moral persuasion, a sort of robotic slave to a meme that you are unable to evaluate." Like the other famous New Atheists, Harris believes there "most certainly" is a conflict between science and religion, and that religion is not only inferior to scientific knowledge and incompatible with it, but also corrosive, and irrational. "The conflict between religion and science is inherent and (very nearly) zero sum," writes Harris.[10] "The success of science often comes at the expense of religious dogma; the maintenance of religious dogma always comes at the expense of science." Harris calls atheism the "plain truth" and is frustrated that he cannot discuss atheism "without offending 90 percent of the population"—a statement that does little to dispel the idea that most scientists are atheists, aloof elitists unconcerned with the common good.[11]

We could go on with colorful quotes.

New Atheist claims about an intrinsic conflict between science and religion have received criticism from a variety of scholars and scientists who have tried to push back on their ideas and thought paradigms. As historian Ronald Numbers and scientist Jeff Hardin[12] write in their edited volume *The Warfare Between Science and Religion: The Idea that Wouldn't Die,* from time to time the New Atheists seem to foster an inaccurate reading of history, claiming that "religion has *always* impeded the progress of science," a claim that Numbers, in particular, has done much to dispel. Their work has indeed been so popular that—even for those New Atheists who are not scientists—New Atheism goes hand in hand with science. This means that many members of the public think that New Atheism among scientists is typical of all atheist scientists.

Prominent religious leaders have also responded to the many publications and public statements of the New Atheists. They criticize the New Atheists' logic, philosophical approach, and methods of public engagement. For example, Tim Keller,[13] former pastor and *New York Times* bestselling author, thinks that the drawbacks of New Atheism are not theological, but civil: "One bad thing about the New Atheist books is they weren't just saying that religion is wrong, they were actually saying that even respect for religion is wrong and that we shouldn't even be courteous and respectful to religious believers, but we really just need to get rid of it all. . . . I think that's a recipe for disaster. That certainly doesn't bring about civil discourse at all." The late Jonathan Sacks,[14] author of *The Great Partnership: Science, Religion, and the Search for Meaning*, and former Chief Rabbi of the British Commonwealth, also found flaws in New Atheist thinking: "I really admire Richard Dawkins' work within his field, but when he moves beyond his field, he must understand that we may feel that he's talking on a subject in which he lacks expertise. Science takes things apart to see how they work. Religion puts them together to see what they mean. And I think the people who spend their lives taking things apart to see how they work sometimes find it difficult to understand the people who put things together to see what they mean."

Science on one side, religion on the other—psychologists will tell us that this kind of binary thinking is cognitively appealing, especially in relation to topics that evoke a strong emotional reaction. Some scholars believe it may allow us to conserve mental energy by helping us simplify information in the world around us. "Reducing complex phenomena or choice to a binary set of alternatives is part of human nature, a fundamental mechanism deeply engraved in our nervous tissue and passed on from generation to generation for our survival," explain organizational psychologists Jack Wood and Gianpiero Petriglieri. "But it can continue to exert an archaic hold on us beyond its usefulness if it prevents us from looking beyond the polarity of 'opposites.'"[15] In short, such binary thinking can lead to stereotypes that have consequences. So when we think of or engage with scientists, especially atheist scientists, we are apt to make the snap judgment that they are against religion and religious people too (because the conflict narrative is the one that is easiest to believe).

The idea that all scientists are atheists who are against religion is a modern myth that drives polarization in society and even keeps certain groups (like women, Black and brown Christians, and the religious more broadly) out of science. Our research shows that there are varieties of atheism among scientists *and that not all atheist scientists see conflict between science and religion.*

If you are a sociologist, one way you confront a myth is with empirically backed data and fieldwork to gather the stories that accurately represent groups of people. During six years of research, we conducted surveys with 1,293 atheist scientists—both biologists and physicists in the U.S. and U.K.[16]—and then interviews with 81 of them as part of a larger study of how scientists in different national contexts approach religion. (At times in this book, we also present data from other surveys, both our own and some that are publicly accessible; we talk more about the data behind the claims in this book in a short methodological appendix.) Naturally, some of these atheist scientists share the views of the New Atheists. More often than not, however, atheist scientists in the U.S. and U.K. have no interaction with religion but are in no way against religion or religious people. A number of these scientists simply do not think or care about religion. But we also discovered atheist scientists with a robust spirituality, and those who have actively made choices that bring elements of religion into their lives.

The New Atheists have performed an important role: They have given a voice to a historically marginalized group of nonreligious individuals, and made it more acceptable to be an atheist, while also raising the profile of the relationship between science and religion. For example, Jesse M. Smith,[17] who focuses on the collective identity work of contemporary U.S. atheists, says that, "only within the last decade have explicitly atheist groups proliferated and become conspicuous." Yet the New Atheists are selling a particular kind of demographically narrow atheism (*notice that all of the voices of public New Atheists we quoted earlier are white men*) and because of the reach of their influence through media and engagement, many people believe they are representative of what atheism is and what atheist scientists believe about religion and religious believers—even when the broader scientific community is characterized by differing stances on these issues.[18]

It's so important to tell the true story of atheists in science: the origins of their secularity, the variation in what it means to be an atheist, their understanding of the limits of science, and the ways they construct meaning in the absence of belief.

Why Focus on Atheism among Scientists?

Francis Collins is one of the most successful scientists working today. He is the current director of the National Institutes of Health (NIH) and the former director of the Human Genome Project. He is also an outspoken Christian. Yet in the U.S. survey Elaine conducted with Christopher Scheitle for their book *Religion Vs. Science: What Religious People Really Think*, they found that both religious and nonreligious Americans were five times more likely to have heard of Richard Dawkins than Francis Collins.

Dawkins and his fellow New Atheists have used their prominence and influence in the U.S. and U.K. to successfully spread the idea that science and religion are inherently incompatible adversaries—and that this is what almost all other scientists also, or ought to, believe. A particular kind of atheist scientist has thus become the public face of the science community. As a result, many members of the public think that all scientists are atheists and all atheist scientists are New Atheists, militantly against religion and religious people.

Wrong-headed beliefs about atheist scientists are consequential. For example, the idea that all scientists are atheists who are hostile to religion makes it harder for members of religious groups to interact with scientists and engage with science. In interviews Elaine conducted for a previous book, she found that a significant minority of religious people, especially conservative Christians, Black and brown Christians, and women, think that scientists—and in particular atheist scientists—are against them. "I think that [scientists] would be more apt to find a bone in the ground and say 'this is the missing link, this is what proves evolution, we can finally shut up those rotten Christians,'" one woman who attends a largely Black evangelical Christian congregation[19] told Elaine. In other work, we found that Catholics, evangelical and mainline Protestants, Jews, and other religious groups are more hesitant than the religiously unaffiliated to recommend that their children pursue careers in physics, biology, and engineering.[20] Some religious respondents told us that their perception that all scientists are a particular kind of atheist scientist is partially responsible for keeping them—some of whom already feel marginalized when it comes to science because of their gender and their race—from entering science. For example, the U.S. scientific workforce is overpopulated with white men. Women, as well as Black and Latinx Americans, are overrepresented in Christian communities. If Christians don't feel comfortable working with scientists or do not believe they can and will be accepted by scientists, they are unlikely to pursue scientific careers, and racial and gender minorities are more likely to remain underrepresented in science.

Finally, the implications of polarization between some religious communities and scientists may be most acute in times of crisis. While a majority of religious communities are no different from nonreligious individuals in terms of their attitudes toward and knowledge of science, some very conservative religious groups respond in kind to the hostility of New Atheists. This impacts how we as a society respond to enduring problems such as climate change or the COVID-19 pandemic. For example, the Cornwall Alliance—a network of evangelical Christian scholars—promotes the idea that environmental advocates push "anti-Christian environmental views" that amount to dangerous extremism.[21] Such distrust has been on display in the U.S. response to the coronavirus.

For example, some religious leaders disregarded the danger of the pandemic and refused to pause religious services when the need for social distancing was most acute; such views received media coverage even though most religious groups followed guidelines to protect against the spread of the disease.[22] Some felt betrayed by epidemiologists who told them that they could not hold a dying family member's hand or celebrate Easter.[23] Assumptions about what scientists actually think about religion—especially assumptions regarding the hostility of scientists—can only further polarize two populations that, in the interest of the public good, must work together in times of uncertainty.

Why Focus on the U.S. and U.K.?

We studied atheist scientists in the U.S. and U.K. because it is in these two nations that the New Atheists rose to prominence and because the public perception of the religiosity of scientists matters in particularly important ways in these two societies. The U.S. is an especially Christian nation. Members of conservative Protestant traditions, such as evangelicals, account for more than one-third of the population, even as non-Christian faiths, especially Islam and Hinduism, continue to rise. The U.K. is also experiencing greater religious diversity—a decrease in the Christian population and an increase in the proportion of individuals who belong to non-Christian religious traditions as a result of migration trends.[24] The proportion of the U.K. population that belongs to non-Christian religions has grown from 2 percent in 1983 to 6 percent in 2012.[25] From 2004 to 2010, both Hinduism and Islam grew by about 40 percent. The influx of Muslims to the U.K., in particular, has also resulted in heightened public awareness of the intersection of religion and science in public life, as well as new religious narratives and practices that have an impact on the relationship between religion and science in public discourse and the scientific workplace.[26]

Although being an atheist is still controversial, especially in Western contexts, atheism is becoming increasingly visible. For example, although not all of them are atheists, the number of "nones," individuals with no religious affiliation, is increasing in both the U.S. and U.K., and within those groups, those who identify as atheists are increasing in number. In the U.S., religious "nones" increased from 16 percent in 2007 to 23 percent in 2014.[27] And there is evidence that atheism is growing.[28] Data from the Pew Research Center indicate that the percentage of the American public who self-identify as atheists doubled from 1.6 in 2007 to 3.1 in 2014.[29] A Gallup poll indicates that atheists represent 11 percent of the U.S. public. These figures may also underestimate actual levels of atheism. (Drawing on a statistical technique that indirectly captures socially undesirable attributes such as atheism, University of Kentucky psychologists Will Gervais and Maxine Najle analyzed two surveys of U.S. adults and estimate that about 26 percent of Americans today do not believe in God.)[30]

Public sentiments toward atheists are beginning to change in the U.S. as well.[31] Over recent years, the percentage of Americans who say it is not necessary to believe in God to be moral and have good values has gone up from 49 percent to 56 percent, according to the Pew Research Center. Yet, Pew still finds that Americans like atheists *less* than members of most major religious groups, and research by sociologist Ryan Cragun and colleagues found that 41 percent of self-identifying atheists had experienced some form of discrimination during the past five years.[32]

In the U.K., the Anglican Church still asserts considerable cultural influence,[33] yet membership in the Church has long been on the decline. Many Brits now "believe without belonging," to use the phrase coined by the prominent U.K. sociologist Grace Davie.[34] In addition, more than half the population now identifies as nonreligious. The British Social Attitudes Survey shows that the proportion of individuals who report no religion increased from 43.4 percent in 2003 to 50.6 percent in 2013. The push for secularism in the U.K. is not a new phenomenon, but in recent years, New Atheists have made significant attempts to secularize society, criticizing religion and arguing that it is incompatible with science and has no place in the public sphere;[35] both Dawkins and Harris have been outspoken in their critiques of Islam in particular. Debates regarding the proper relationship between science and religion now feature prominently in U.K. public life. In short, the Western world is seeing a steady rise in nonreligion and a growth of atheism.

This book is about the varieties of atheism in science. In some ways the title is a play on *The Varieties of Religious Experience* written in 1905 by the psychologist William James, who argued that the religious tradition one belonged to was less important than the varieties of psychological and spiritual experience religious individuals have in their own terms. Our foray into atheism perhaps builds on that approach. Until now, what atheist scientists actually think about secularity and religion has primarily been understood through the narratives of vocal celebrity scientists. And while their work has had a powerful impact by giving voice to nonreligious individuals and articulating the New Atheist platform, it nevertheless puts forth a narrow view of atheist scientists—who represent the vanguard of atheism in society. Here we use nationally representative data and in-depth interviews to identify the origins of atheism in scientists' lives, the variety of atheist identities that exist in science, the meaning systems that these scientists espouse, and their views of the limits of what science can explain, all in the terms of atheist scientists themselves.

Chapter 2, "Tried and Found Wanting," examines religion and secularity in the life course of atheist scientists. To do so, we use survey data to identify exposure to religion (or lack thereof) during the adolescence of contemporary atheist scientists. We also examine whether exposure to science played a role in atheists' transitions away from the religious communities of their youth. The chapter also draws on interview data to consider how the scientists explain these transitions in their own terms, including the experiences of individuals who grew up in atheist families.

Having identified the origins of atheism in scientists' lives, the next three chapters take us into the three core atheist identities we found in our analysis of scientists in the U.S. and U.K. In Chapter 3, we encounter the *modernist atheist* scientists whose identities are—in principle—most like the New Atheists. They do not believe in God, they are not spiritual, and they rarely interact with religious individuals or consume religious culture. This group does bear some resemblance to the New Atheists. They are more likely than any other atheist scientists, for example, to describe the relationship between science and religion as one of conflict. Nevertheless, while many of these scientists are concerned that religion may be harmful to science and society, we also encounter scientists in this group who are simply indifferent to or even appreciate religion. What is more, we find that many modernists are concerned that the engagement style of New Atheists such as Dawkins in the public sphere is unproductive and risks presenting the public with a biased view of science and scientists.

In Chapter 4, "Ties That Bind," we discuss *culturally religious* atheist scientists who have sustained patterns of interaction with religious individuals and organizations. This chapter takes us into the lives of atheists whose choices and actions depart most from the picture of atheism we see among New Atheists. Here we consider scientists who have married religious individuals, who no longer believe in God but belong to a religious community, who send their children to religious schools, and who pursue other forms of connection to religion. Our analysis will explain the religious traditions to which some culturally religious atheist scientists continue to belong, how they make sense of such connections, and the benefits they claim from such ties.

In Chapter 5, we talk about frameworks of those scientists who fit the category of "spiritual atheists." These atheists tend to define spirituality as a deep sense of awe combined with moral implications, and they see it as compatible with science. This seems to be a different view of spirituality than that held by the general population of the U.K. and U.S. This chapter delves into these unique perspectives and explores the many ways that scientists can use spirituality in their lives, from practical efforts to run a "kind lab" or justify humanitarian efforts, to conceptual attempts to wrangle with death, dying, and the meaning of life.

Having established the three main atheist identities we observe in science, we then turn to two broad themes: the limits of science and meaning systems among atheist scientists. Chapter 6 examines atheists' rhetoric of science with two questions in mind: How do they distinguish science and religion as ways of knowing? And, if science is the only legitimate way of understanding the world, are there limits to what science can tell us about the world around us? As we will see, rejecting religion as a way of understanding the world is central to what it means to be an atheist. Nevertheless, while some atheist scientists believe there are no limits to what science can explain—including branches of knowledge beyond physical science and emotions such as joy and sorrow—other atheists believe human cognition and technology will always preclude science's ability to explain reality.

In Chapter 7, "How Atheist Scientists Approach Meaning and Morality," we examine how atheist scientists grapple with life's big questions. Do they find questions about meaning and purpose important? If they believe only science can explain the world around us, where do they create meaning in their day-to-day lives? And, what do their approaches to meaning and purpose tell us about morality among atheist scientists? The most prominent narratives we heard among all atheists were that there is no meaning or that questions of meaning cannot be answered. Still, even as atheist scientists characterize the world as random, purposeless, and godless, they construct a progress-oriented meaning system in which science leads to advancements that transform and improve the world. We then turn to questions of morality and—drawing on our broader survey data—find almost no differences in moral commitments of religious and nonreligious scientists in the U.S. and U.K. While some in the public sphere view religion as a moral safeguard, atheist scientists view morality as emerging from a rational examination of how to improve well-being and the world around us.

In Chapter 8, "From Rhetoric to Reality: Why Religious Believers Should Give Atheist Scientists a Chance," we consider how the New Atheist monopoly on the image of the atheist scientist is consequential for religious and scientific communities. If science wants to attract more women and minorities to its community, and enhance public trust, it needs to ensure these groups have a more accurate view of how atheist scientists think about and approach religion and those who are religious. We provide tools for dialog between these seemingly disparate groups.

As we will demonstrate in this book, the popular conception of the atheist in science is misinformed. Most members of the general public do not know that atheist scientists are by and large *not* hostile to religion. Also, atheist scientists and religious communities may disagree about the origins of creation or the forces that shape the natural world, but many of them share a fascination and awe of the world, a sense of meaning and purpose, and a desire to explain something larger than themselves. In some cases, atheist scientists are sitting in the pews, participating in the same services as religious individuals, or engaging in religious rituals at home with their families. This book highlights what these scientists really think about religion, spirituality, and the limits of science to shine a light on the true nature of atheism in the scientific community.

It is now time to broaden the dialog, making room for a wider variety of atheist scientists, along with their ideas, narratives, lived experiences, and approaches to atheism. That is what we hope to do on these pages. We want to tell the true and real story of atheism in science, not based on the loudest voices but rather the soundest data. It is a story that needs to be told.

2

"Tried and Found Wanting"

How Atheist Scientists Explain Religious Transitions

On a spring day in April a few years back,[1] we travelled to an elite university in the Northeastern United States for an interview.[2] We were meeting Emily, a Ph.D. student in the physics department. She wore jeans, a T-shirt, and tennis shoes and had fiery red hair and rectangular glasses. Her research focuses on galaxies.

Emily was raised a Christian and attended church and Sunday school throughout her young life. When she took AP Physics and other science classes in high school, however, she began to question what she had learned about the faith and what church leaders had told her, leading to doubts. "I actually am confirmed in the Presbyterian Church and if you look at my confirmation statement you can sort of see it happening already," she said. "It's not so much, 'I believe that Jesus lives and that God saves us,' and that sort of thing. It's more like, 'I think that this is a framework for morality. I think there are useful lessons to be learned from religion and that the way that the Bible tells us to behave is a good thing. I believe that God exists, but maybe not exactly how the Bible works.'"

Even though she was raised in a religious environment and at one point found Christianity personally important, Emily eventually transitioned away from religion and now identifies as an atheist. She said several factors influenced her decision to transition away from faith. Her scientific learning played a part in her questioning of Christianity, but she also had people around her

Varieties of Atheism in Science. Elaine Howard Ecklund and David R. Johnson, Oxford University Press.
© Oxford University Press 2021. DOI: 10.1093/oso/9780197539163.003.0002

who shared her doubts about faith and who were willing to engage in critical conversations. She told us about one friend, in particular:

> He often challenged my views on religion. So I presented him with this picture of, "I believe in God. I believe he set up the physical laws and the initial state of the universe and then said let's go." He introduced me to this sort of, I don't know what to call it, but it's the "God of the gaps." So basically whatever you don't know, stick God in there. At first I was just angry and I didn't listen to him, but over the years I have thought about that more and more and decided that whether or not God exists is not useful for me as a scientist to think about him. Because I need to pretend he doesn't exist because that's what science does. It says let's find the provable things, and God is not provable.

Emily ultimately gave up religion, despite having religious convictions as a teenager. A longstanding assumption is that individuals like Emily, who were raised in a religious tradition, abandon their faith as a result of exposure to scientific knowledge. But as we see, her journey away from faith involved not only science education, but also social networks and personal reflection. As her pathway suggests, a number of factors may influence why someone whose childhood socialization emphasized religious faith ultimately rejects the identity, practices, and beliefs they once embraced. Sociologists have long studied such transitions among the general public and their research points to how life events such as getting married, having children, or particular educational trajectories lead individuals toward or away from religious faith.[3] Many scholars assert that educational attainment is important to why individuals become atheists, but sociologists Damon Maryl and Jeremy Uecker find that the influence of higher education pales in comparison to social networks—the influence of parents, friends, and religious communities themselves.[4]

Researchers have thought for a long time that an increase in trust in science and the proliferation of scientific institutions secularizes societies broadly. Support for the idea that *science secularizes* emerged in part from research on the religiosity of scientists themselves. Much of the existing popular research on the religious lives of scientists tends to find that scientists are more likely to be atheists than the general population and infers from such findings that lack of belief in God is specifically because scientists' expertise helps them evaluate religion more critically than the average person. Such investigations into the religious lives of scientists first emerged with the work of the psychologist James Henry Leuba.[5] What he found, and what many have later tried to explain, was that U.S. scientists are less likely than the general public to believe in God. In a 1960s study, sociologist Rodney Stark found that more than one quarter of U.S. graduate students in the sciences claimed no religious affiliation, compared with just 3 percent of the public at the time.[6]

One idea, in particular, that has fueled the assumption that science secularizes is that the "best scientists" are the least religious, suggesting that the absence or abandonment of religious faith is a prerequisite for eminence in science. Some researchers have reasoned that the most elite scientists—like those in the National Academy of Sciences in the U.S. or Fellows of the Royal Society in the U.K.—would likely have the highest and deepest knowledge of science and therefore be especially likely to reject religion in favor of science.[7] When Leuba collected data on the religiosity of scientists during the early 20th century, his study revealed that scientists at elite institutions were less likely to believe in God when compared with scientists at less elite institutions. When Edward Larson and Larry Witham later replicated Leuba's early 1900s study, using data on members of the National Academy of Sciences—considered the most elite body of U.S. scientists—they found that religious disbelief was most common among these scientists.[8] They found that only 7 percent of members of the National Academy of Sciences reported belief in a personal god.[9] Researchers have found similar evidence among the most eminent scientists in Great Britain. In a recent survey of the Fellows of the Royal Society of London, only 5 percent of scientists reported belief in a personal god.[10]

These studies, of course, select narrowly upon an exceedingly small subset of scientists. Scholars also argue that part of this perceived incompatibility between science and religion may be due to social networks in elite research institutions rather than just knowledge about science itself. Elite institutions, and the social networks they facilitate, may indeed be as relevant or more relevant than learning more about science itself in shaping the religious attitudes and commitments of scientists. For example, sociologist Randall Collins argues that the academy plays a central role in shaping ideas about the development of worldviews. It is within university settings that academics form the kind of intimate social networks that help them become leaders in the transformation of culture. Elite university scientists, then, also have an important role in knowledge creation and institutional change, because they provide scientific training to future societal leaders. But these same connections may make scientists and the students who study with them susceptible to cultural pressures to be irreligious as part of being a scientist.

More importantly, the idea that science secularizes is complicated by considerable evidence that many scientists remain religious. Indeed, in our previous work we have found that—across multiple national contexts—more scientists are religious than most people assume and that the most common view on the relationship between religion and science among academic scientists is one of independence.[11] In the tradition of the late paleontologist Stephen Jay Gould's[12] idea of "non-overlapping magisteria," faith and science are seen as different ways of knowing about different aspects of the world. So then, many scientists believe they do not need to abandon religious convictions to work as a scientist: It is completely possible to do their scientific work while keeping their religious affiliations and propensities separate from that work (although many scientists do indeed integrate their scientific work with their faith). Elaine also found in her previous work that U.S. scientists at elite academic institutions are more likely than the U.S. public to be raised in households without religion, thereby suggesting that those who are already without religion might be self-selecting into science.

If we examine the atheist scientists in our study, we see that 47 percent in the U.S. and 35 percent in the U.K. were not always atheists. These individuals have past connections to religion, often very intense ones; they were born into religious families, raised in a religious faith, and asked to believe religious ideas and doctrines. If exposure to science is as influential on becoming an atheist as many scholars assert, we should certainly expect to hear scientists talk about their atheism as coming about mainly because of science. In what follows, we consider pathways toward atheism among scientists in the U.S. and U.K., the rationales behind such transitions, and the role of socialization to science. As we will see, exposure to science and scientific training indeed contribute to becoming an atheist, *but are less influential* than many assume.

Religious Transitions of Atheist Scientists

While the factors driving individual secularization are still contested, recent research has highlighted the importance of trust in science for "achieving" an atheist or nonreligious identity. Jesse Smith[13] argues that science provides "an effective, institutionally grounded meaning structure" for avowed atheists. Other research has discussed the important role of family in individual-level secularization.

When we examine together the U.S. and U.K. atheist scientists who participated in our survey, we find that 53 percent have never had a religious affiliation, 27 percent were affiliated with

Christianity at age 16 and are not currently affiliated with a religion, and 7 percent were affiliated with another religion at age 16 and are not currently affiliated with any religion. In addition, 11 percent of these scientists had a religious affiliation at age 16 and maintain that affiliation today, while expressing that they do not believe in God. Only a small minority, 3 percent, did not have a religious affiliation at age 16 but do affiliate with a religious tradition now. Many scientists in our study reported transitioning away from religion because of science, secular family members, or bad experiences with religion.

Transitioned Away from Religion Because of Science

When we began this work, we anticipated that the scientists we encountered would report leaving religion because things they learned in science conflicted with their faith. That is, we thought that prolonged exposure to the undergirding philosophy, training, and work of science in a professional setting would provoke a break with religion. However, we found that fewer than half of atheist scientists who were exposed to religion as children think science played a role in their transition away from religious belief. In the U.S., 46 percent of atheist scientists reported that their scientific training and knowledge made them less religious, while 54 percent told us science had no effect on their religiosity. In the U.K., 40 percent of atheist scientists said their scientific training and knowledge made them less religious, while 60 percent reported that science had no effect on their religiosity. In both countries, we see that science encouraged a secular path for many, but not most, individuals who grew up in religious households.

For those scientists who did attribute their transition away from religion to exposure to science, we learned in interviews that the methodologies, positivism, and empirical evidence–based approach of science fascinated them. When they viewed their faith through the same lens of the empiricism of science, however, religion came up wanting on empirical grounds. Others were more resolutely religious and transitioned away over a long period of time as they learned more about science; they would eventually give up their faith, but only after much wrestling and thought.

For U.S. and U.K. atheist scientists, "religion" more or less means "Christianity." Among the U.S. atheist scientists in our study, 52 percent were not part of any religion at age 16, 15 percent were Roman Catholic, 18 percent were Protestant (either Mainline or Evangelical Protestant Christians), and about 15 percent fall into other categories. Among the U.K. atheist scientists, 65 percent said they did not belong to a religious tradition at age 16, but 13 percent were Roman Catholic and 17 percent were Protestant Christians (largely Anglican) at that age. And roughly 13 percent were in other categories. While the transition to atheism was similar in many ways for U.K. and U.S. scientists, key differences did emerge in our interviews. Scientists from the U.K. felt that they were breaking from *institutional* Christianity, namely that of the Anglican Church, while U.S. scientists, particularly those who grew up in the South or rural communities, felt that they were breaking from *cultural* Christianity.

U.K. Scientists

For the scientists who were raised in England, there was a particular kind of tension between science and religion distinctly related to the Anglican Church, a state church that is very influential, even though religious participation in the U.K. is quite low. The church's influence includes state-funded schools, which have historically included religious education and daily acts of collective worship, even as rules governing such practices have become less stringent over time. References to religious culture in state schools frequently emerged in the accounts of scientists we interviewed in the U.K. One biologist[14] we spoke with concisely captured the type of narrative we heard from many scientists there: "I went to a church school and obviously . . . being a teenager you don't really want to fit in particularly well because you have a state religion essentially, and I was at a school with a religious ethos. I guess I wasn't very rebellious as a teenager, but . . . the sort of separating yourself from your immediate surroundings, which is what happens when you form your own identity, I think [the religious transition] was wrapped up in that." For these scientists, Christianity figured prominently in early educational training, meaning that becoming an atheist involved rejecting one of the key "official" narratives of religious culture.

Unsurprisingly, the influence of exposure to science figured prominently in the narratives U.K. atheists used to explain the religious shifts in their lives. Our survey found that 40 percent of U.K. atheist scientists indicate that scientific knowledge and training made them less religious. In particular, nearly 29 percent said science made them "much less" religious, while about 12 percent indicated their training made them "slightly less" religious. Many scientists described the process as a gradual transition away from religion, often involving attempts to reconcile a scientific understanding of the world with a religious one. One scientist from the U.K.,[15] for example, told us how he tried to fit his religious views and his scientific views together, but could not quite make it work. "I was just trying to accommodate my naïve religious views of the time with what I was learning from science and I saw that actually the two things don't go together very well," he said. "It's not just about the notions; it's about the methodology that in religion you should have faith . . . without any proof of it essentially while science it's the opposite, you know. You shouldn't believe in things without evidence."

While scientists in the U.K. often attributed becoming an atheist to their scientific education, their narratives rarely depicted a clear or linear process. More often than not, these individuals formulated their own critiques of a faith-based view of the world as adolescents, prior to sustained exposure to scientific understandings. Consider, for example, a U.K. graduate student in physics[16] who told us that she transitioned away from religion at a young age because of reason and science. For her, the transition to atheism was as natural as rejecting mythological figures in folklore. In her sense of things, science helped her distance from the myths of religion. She said that "when you're a child and . . . you're exposed to the stories, initially it's like Father Christmas: you believe it. But when you start asking about how he manages to get down the chimney, and you start finding the answers are nonsense, then it doesn't take very long not to believe it." Rejecting what she saw as religious myths and dismissing the explanations that accompany them did not require any education or training in science. Nevertheless, this graduate student[17] attributed her transition to atheism to a budding interest in science, specifically astronomy; in her view, "as a very small child I believed in the way in which small children do. But as I began to question explanations, then [I realize now that] even a small child can see the inconsistencies between religious explanation and reality."

Similarly, a biologist in the U.K.[18] knew from an early age that religious explanations were not rigorous enough for her, despite her parents' devout Muslim faith. She explained, "I just don't believe in the stuff that they say about religion. Don't believe—I mean I was actually studying the Koran at the age of thirteen and I said to my dad, 'I don't think I believe in God,' and he was—I think he talked to me about it. He was very religious himself and, yeah, I just can't believe in those stories, so—and I've never found anything else."

The interviews, of course, provide retrospective accounts, meaning the sequences of events and attributions of influence are subject to the limits and biases of memory. Nevertheless, the narratives we heard from atheist scientists in the U.K. suggest that even when exposure to scientific knowledge is not understood by these scientists as the cause of doubts about religion and their transition away from faith, this exposure often affirmed prior doubts and raised more questions that undermined their faith. The atheist scientists we spoke with in the U.K. often revisit core tensions between religion and science, and many have evolved in their view of this relationship. One of the more typical transitions involves adopting a "hard line" stance that's pro-science and anti-religion, which ultimately softens over time.

An example of this pattern is found in an interview we conducted with a U.K. biologist[19] who summarized his shift to atheism not so much as a response to a struggle with the faith of his youth but more as a reaction to statements by religious individuals in his community who said what he now describes as "profoundly silly, unexamined things." Being around people who did not have a particularly intellectual faith, he said, at the same time as he was learning more about science, "gradually chipped away at what I thought I believed. [*five second pause*] And I thought—I mean, typical teenager—I thought there were people using religion to justify some basically untenable viewpoints, and it's a very easy way of justifying anything, at least is what I thought at the time." However, he now thinks that these issues are not quite so clear-cut, explaining, "Obviously that was something that was played out as I lost my faith, and I think I probably had a much simpler idea of the difference between them [science and religion] than I do now, obviously." This scientist no longer sees religious people in the "simplistic" way he did when he first became an atheist, instead reflecting, "I think there's an underlying philosophical difference [between science and religion], which is more difficult to reconcile. But I think at that level, actually you can't really say that one is better than the other, other than that people who are scientifically advanced tend to have more—have more stuff essentially is what it comes down to."

U.S. Scientists

The U.S. is a much more religious country than the U.K. While the U.S. is officially secular, religious communities are highly influential across various realms of public life to such an extent that atheists are a relatively silent and marginalized group. In short, the country is characterized by a pervasive cultural Christianity. Consequently, many U.S. scientists received much more intensive exposure to religion than did atheist scientists in the U.K., though that exposure was oriented around pressure toward personal belief rather than adherence to a state church.

Many of the atheist scientists we met in the United States told us they grew up religious in regions of the country where Christianity is the predominant worldview, and they left Christianity after they went to college. We found that college became a key context of cultural and religious transformation for many of these individuals. College allowed them to escape the pervasive cultural oversight of their childhood communities and meet people with different views and experiences, and thus provided them the freedom to doubt, examine, and ultimately leave their faith.

U.S. scientists were only slightly more likely than scientists in the U.K. to report that scientific knowledge and training had an impact on their faith. For 45 percent of respondents, exposure to science made them less religious (compared with 40 percent in the U.K.). Most scientists indicated that science played a role, but the narratives consistently suggested that scientific knowledge was one of many factors. A subset of U.S. atheist scientists viewed exposure to scientific knowledge as the primary cause of becoming an atheist, though they also mentioned other influences, such as changing social networks. As one example indicative of this pattern, consider a graduate student in biology[20] we met who turned away from religion while she was in college. She said she was influenced by the people she met and experiences she had there. She found a group of friends who were all questioning their Christianity, she said, and they came to a point when they were "realizing we weren't so into [Christianity]." She also told us about the influence her professors had on her as she learned to think about science, religion, and the process of prioritizing evidence. She had professors who "had a view of life kind of like 'science first,' not that science is more important than religion, but like what you find out from science is what's true," she said. They told her that when science conflicts with the Bible, "you re-interpret the Bible because that's not science, that's like some religious document, that changes," she remembered, and she began to question a lot of the things she was taught as a child. "I think as I learned more about that kind of viewpoint, and also how a lot of stuff that I was taught as a kid . . . that's just not true of science—like, I guess like dumb things about evolution, and science—I was like oh, all that stuff that I was taught, just isn't true," she said. For many aspirant scientists, and other students more broadly, college exposes them to new ways of understanding the world. These new ways of understanding often mean having to reconcile old and new worldviews, and sometimes choosing to privilege science over religion.

The qualitative data from our interviews with U.S. atheist scientists often suggests, like it did with U.K. atheist scientists, that science education and training probably reinforced rather than started their transition to atheism. For example, some of our conversations resembled one we had with a graduate student in physics[21] who was raised an evangelical Christian but said she "never really fit in" at church. The "epiphany moment for me was when . . . we were talking about how to convert people to Christianity," she said. "I remember asking, 'Well, if they don't believe me, how do I prove it to them?' Sounds like a pretty science thing, right?" she recalled. "And my Christian mentors said, 'You don't! You just show them by your faith.' And I'm like, that doesn't exist, I'm done." Scientific knowledge is not what led this young physicist to determine that she did not "fit in" among her religious community, but it did affirm this feeling and provide her with a rationale for embracing atheism.

Bad Experiences with Religion

As we have seen, exposure to science was a factor, even if not the precipitating factor, in moving many of the atheist scientists we met away from the religious faith of their youth. But putting too much weight on scientific training as the most influential factor in developing atheism among scientists obscures a possibly more important influence on why individuals—and the scientists in our study in particular—become atheists: personal experiences within their religious communities. A significant minority of the atheist scientists we studied attributed their transition to atheism to bad experiences with religion.

For some of these scientists, the tensions were doctrinal. One scientist who was raised Catholic,[22] for example, told us that her doubts about religion could be traced back to her earliest experiences with her religious tradition rather than to her experiences with science. She appreciated the communal aspects of religion, she said, but from the very beginning the stories of her faith struck her as harsh. "I think I liked the experience of being in a church, but . . . from the very beginning, I had some misgivings about the stories, and I would actually ask my grandmother a lot about that, and she didn't really have good answers. . . . So this was just not very satisfying to me," she explained. "I was opposed to the more cruel parts of the Bible, so the story [of] Jesus being nailed to the cross was just something that I was scandalized about, how that could happen to people."

Another scientist[23] told us how her move toward atheism started with rejecting the claims of religion as ridiculous. When she was starting high school, she was "going to Sunday School," she said, "but I wasn't terribly keen on going to church and Sunday School and I think there was a subconscious questioning going on and I wasn't very convinced about the whole story." She turned to a high school history teacher, who told her that religion had been invented by man, and "that came as a bombshell to me and I felt—I remember, yeah, it kind of again impacted things that were going on subconsciously and it made me think: yeah, this is all kind of [a] not true story, fairy tale." At this point in her life, she experienced a transition away from religion, like many of the other atheist scientists we've introduced. In her teenage years, someone tried to "rescue" her from atheism, she said, and they "talked me into the wing of this very conservative faction of the Church.....I think [their] particular view of the world was incompatible with a lot of what was going on around me and a healthy development ... of an adolescent and ... ensured ... many, many years of ... re-questioning again and finding a way out."

Other atheist scientists locate the origins of their atheism in difficult experiences with religious leaders or members of the faith community of their youth. One biology professor[24] discussed growing up in a family in which her father was an atheist who sometimes attended church, and her mother was "a very religious Catholic." Because of her mother, she was brought up Catholic and sent to a Catholic day school run by nuns at a convent, which she began to resent at an early age because of the harshness with which the students were treated:

I was four years old when I went to school and we were being trained not to tell lies, and if you did something naughty, you were beaten quite savagely. . . . Young children used to wet their knickers. It was a very common occurrence, and if somebody wet their knickers, they were beaten. I mean that's not what you expect is it?

She talked about the abuse, shame and hurt she experienced at the convent, where she observed how those who purported to serve God contradicted their beliefs daily, and having to attend mass every single day, this professor[25] concluded that religious people are hypocritical. The hypocrisy she saw concerned abuse, shame, and hurt that she and others experienced over actions as small as alleviating a little bit of discomfort. By way of example, she told us this story:

> We were going to be christened, a christening/confirmation thing, and we were practicing taking the holy communion, which is where they put the bit of bread on your mouth and then you're meant to swallow it. There's a little rice cakey thing, you know, this little, very dry paper, and I got mine stuck on the roof of my mouth. And I can remember it to this day. It got stuck right on the roof of my mouth and it was dry. . . . I put my finger in, just to flick it down. Unfortunately, I flicked it out of my mouth [*laughs*] and I was savagely beaten for that, because it wasn't blessed by the priest, because it's not the body of Christ until it's blessed, but it went on the floor and the nun said that could have been the body of Christ, and I said, but it wasn't, because the priest is not here. . . . I can remember the unfairness of it and then I can remember you know, hands out, big wooden ruler, whack, whack, whack, so you had big welts on the backs of your knuckles. Really painful, and I thought then: if this is what religion is, and this is the way religious people behave, I don't want anything to do with that. And it just escalated from there, the things that they used to get up to, these nuns. So that finished me, I'm afraid, for religion, and then I learned more and more about what the Catholic Church was doing, and what the Catholic priests have done. You know, I don't really need to embellish that at all, and I have many personal experiences, which I mean you can probably guess.

Some of the atheist scientists we spoke with told stories of negative experiences they had with other members of their faith community. One U.S. scientist,[26] for example, attributed her atheism to bad experiences with other religious individuals in her childhood community, which was a highly religious small town. Her family was religious, but attended services less frequently than other community members, which led some of her peers to recurringly pick on her. "It probably made me a little antagonistic to religion because I was told I was going to hell when I was a kid by classmates, etc., because we didn't go to church," she told us. Another atheist scientist[27] told Elaine about living in a Methodist orphanage during her formative years while her mother was sick, and the difficulty she had getting back to her family because the orphanage, unbeknownst to her, had adopted her. Her "mother did get well," she said, "and it took my parents quite a while to be able to get me back. . . . And I had my 13th birthday there. It was horrible." Not only did she experience separation from her family and fear about her mother's health at a young age, but she also experienced religious individuals who violated her trust, took her from her family, and kept her from going back when she had the opportunity to do so. It was this defining experience that largely turned her away from religion and toward atheism.

Some atheist scientists attributed their transition away from faith to personal experiences with pain and suffering in their own lives or in the world, which led them to question the existence of an omnipotent, benevolent God. Namely, they were unable to resolve what philosophers call "the problem of evil." One biologist[28] recounted how he could not comprehend how God could allow the suffering he had experienced and observed in his early life:

I figure I *was [respondent emphasis]* religious up to the age of 25 maybe. And I'd slowly diminished . . . *[trails off]*. I don't quite know why it did. I would say by the time I was 30 I was a full-blown atheist. . . . My father had a very long prolonged illness when he died, and I think that was really not the only thing, but that was the straw that broke the camel's back. . . . That degree of suffering, yeah. . . . I think I'd probably lost all virtue of faith before that, anyway. And that was from combinations of reasons . . . a combination of probably the social history of religion . . . the awful, awful problems it's caused in Europe, and was at that time causing in Northern Ireland, for example. Then you wake up one morning and you realize that it's actually just—it's a human invention, really, to facilitate things.

Another biologist[29] discussed how she couldn't reconcile the idea of God as it was presented to her with her life experiences:

> I think two things happened, one was my grandmother died and she was someone that I loved very much and I thought well by my definition, she's now in hell, I wasn't very keen on that idea . . . then we had children and I carried on working, so I'd worked all the way through having children and I used to end up in tears at the end of the services because he [the pastor] would have given a lecture on how mothers should be at home looking after their children and it was incredibly difficult to sit there being told basically that your God doesn't want you to go out to work week after week and that you were being very bad for doing so.

Atheist Childhoods

As we noted earlier in the chapter, 53 percent of the collective group of U.S. and U.K. atheists in our study were unaffiliated with a religious tradition at age 16, so they never made the transition from religion to nonreligion; they were always unaffiliated and generally most did not believe in God. That more than half our study sample fell into this category is noteworthy given the institutional exposure to religion that individuals in the U.K. encounter and the cultural pervasiveness of religion in the U.S. For many of these scientists, their identity as atheists was less their own decision than that of their parents or grandparents. "My family haven't been religious for generations," a U.K. scientist[30] told us. "It's something you learn about in school, but it was never really anything that came in the home very much."

We heard a number of stories like the one from a British graduate student in physics,[31] whose immediate family was not religious. He was exposed to prayer in school, but he never considered himself religious in any sense and never felt any affinity for religion because of his family dynamic. "My family has never been religious. I actually never discussed the matter with them. I can't tell you what—what their beliefs are," he said. "But I do also know there are members of my family who are religious. My [grandmother] in particular, although I never discussed this matter with her again. So in that sense, religion only influenced me from the rest of society." He has no dramatic story about leaving religion; he was simply always an atheist.

A biology student[32] in the U.K. explained that he also grew up in a "quite secular" family and his family's lack of religious affiliation can be traced back to his grandfather:

> [M]y granddad served in the second World War and I think he had . . . quite a tragic time in that. [*laughs*] I think—yeah, so he went through Dunkirk and then he went and served in Japan in the Japanese theater and he got taken prisoner, and . . . on his suffering there, I think he kind of—he denounced religion, so I don't know how religious he was beforehand, but on the basis of that he definitely came to the conclusion that he could not believe in a God and this is something that I think had—a downward effect on my family.

We found that the large majority of atheist scientists who grew up in atheist families do not identify with a religious tradition as adults. Indeed, of the atheist scientists in our study, only 3 percent were unaffiliated with a tradition at age 16 and affiliated with a tradition at the time of the survey.

Growing up atheist, however, does not mean that these scientists were not exposed to questions of religion in their early life. Yet, when such questions did emerge, they were often answered in a way that did not involve or require religion. David interviewed a biologist[33] who, when asked about religion, said he "just didn't feel it necessary." He said he had a "very stable" upbringing, never wanted for anything, and did not endure a crisis that turned him toward God. He also said he "never" experienced a religious shift of any kind. When asked if his family ever talked about religion, he said:

> I wouldn't say it was a topic that came up often around the dinner table, for instance. I mean, I think it was more from my questioning, maybe saying, what do you think about it? And I think because my question was maybe more scientific, maybe through discussion it couldn't be fulfilled. Maybe conversation probably turned more to morals and ethics and I think actually when I think about morals and ethics . . . I'm a strong believer that you don't need to have any religious dogma or scripture to inform you of what is right and wrong. I think it's naturally ingrained within us anyway.

Some scientists who grew up in atheist families allowed themselves the opportunity to explore religion, even in the absence of faith at home. For example, one lecturer in biology,[34] whose parents have always been atheists, explained to us that she attended church occasionally on her own for a period of time, just to see what it was like:

> I had friends at senior school who started going to church and I think being really cynical, I think for some of them, they probably came from religious families and for others it was the cool thing to do because you went, hung about church there, and smoked dope—if we're being strictly honest [*laughs*]—and so I sort of hung around with that crowd a bit. It made me think a bit about whether there was a God or something like that. I attended church services for a bit, which may have even simply been an

act of rebellion against my parents and then I stopped doing it.
[*laughs*]

From listening to the stories of atheist scientists in the U.S. and
U.K., we learned that, in some cases, scientific knowledge and
training did play a primary role in drawing them away from religion and the religious beliefs and practices of their youth. For most,
however, we found that scientific exposure and training had a small
but significant influence alongside other factors, such as negative
experiences with religion or changing social networks, accelerating
a pathway from religion to atheism that was already underway.
Science is clearly integral to "achieving" an atheist identity, as Jesse
Smith notes. But, for those raised with religion, science's core role
may rest more in providing a meaning system for atheist scientists'
current lives, and less in sparking the path away from the religious
roots of their earlier lives. Yet meaning systems, even those organized around highly institutionalized norms and practices as found
in science, are not unequivocal. Subsequent chapters begin to un-
pack this point, demonstrating variety in what it means to be an
atheist scientist.

Varieties of Atheism at a Glance

Before we turn to the distinctive cultural differences in what it
means to be an atheist scientist, it is important to consider the social
differences that characterize who atheists are. Given that the loudest
voices are older white men, we were surprised to find that atheist
scientists span a diverse demographic spectrum, with a significant
portion of them being female, immigrants, or non-white. In Table
2.1, we present a demographic overview of atheist scientists in the
U.S. and U.K. With respect to gender, atheists are predominately
male, but women make up a sizable minority representing one-
third of atheist scientists in both the U.S. and U.K. While the two

Table 2.1 Demographic Overview of Atheist Scientists in the U.S. and U.K.

	Overall (%)	U.S. (%)	U.K. (%)
Gender			
Female	32	32	32
Male	68	68	68
Race			
White	71	67	90
Black	<1	<1	0
Asian	21	24	8
Hispanic	5	7	0
Other	3	2	2
Rank			
Scientists in Training	38	39	34
Early Career Scientists	35	32	46
Senior Scientists	27	29	20
Discipline			
Biology	69	69	71
Physics	31	31	29
Institutional Type			
Elite	53	46	81
Non-Elite	47	54	19
Atheist Category			
Modernist	67	66	73
Spiritual	6	6	6
Cultural	27	28	21

Notes: Based on weighted data. Values exclude nonresponse.

Source: RASIC United States Survey 2015; RASIC United Kingdom Survey 2013

nations mirror one another in terms of gender, atheist scientists in the U.S. are much more diverse when compared to U.K. scientists. In the U.S., two-thirds of atheist scientists are white, one-quarter are Asian, and 7 percent are of Hispanic or Spanish origin. Black atheists, who represent just under 2 percent of atheists in the U.S. public, comprise the smallest racial group at less than 1 percent of U.S. atheist scientists (see the work of humanist-theologian Anthony Pinn for more on the significance of atheism in the Black community).[35] In the U.K., by contrast, atheist scientists are predominately white (90 percent), with Asian scientists making up the second largest group at 8 percent (in both countries, a majority of Asian atheists are non-native).[36]

We also observe diversity in the professional characteristics of atheist scientists. In the U.S., atheist scientists are roughly distributed into thirds across career stages, with doctoral students comprising the largest subset at 39 percent. In the U.K., we see that one in three atheist scientists are in doctoral training, nearly half are in early to mid-career positions (from postdoctoral training to mid-career positions), and one in five are at the upper level of their careers (such as in professor positions). With respect to discipline, atheists are much more likely to be biologists than physicists. In both countries, biologists make up nearly 70 percent of the atheists in our survey, with physicists accounting for one-third of the sample. The main difference in the professional landscape of atheist scientists in these two nations is tied to the prestige of the universities and research institutes where they work.[37] In the U.K., atheists are overwhelmingly (81 percent) located at elite institutions. In the U.S., atheist scientists are more evenly divided across the prestige stratum, with slightly more atheists at non-elite institutions.

Overall, an examination of demographic and professional characteristics of atheist scientists demonstrates relatively few differences between the U.S. and U.K. It also reveals that atheist scientists are *much* more diverse than the celebrity New Atheist scientists. One important dimension of diversity remains that will occupy our focus in the next three chapters of the book: what atheist scientists think about religion. In the U.S. and U.K., we set out to understand who atheist scientists are and what they really think about religion. We asked them not only what they believe about God, but also whether they participate in religious practices and rituals and whether they consider themselves spiritual. Based on our research, we categorize atheist scientists into three groups: *modernist atheists, culturally religious atheists,* and *spiritual atheists* (see Appendix for a detailed explanation of our classification scheme).

Modernist atheist scientists—those who have no spirituality or engagement with religion—represent the largest category, representing three-quarters and two-thirds of atheist scientists in the U.K. and U.S., respectively. Culturally religious atheist scientists constitute the second largest group in the study. These atheist scientists, who make up one-quarter of participants in the U.S. and one-fifth of participants in the U.K., have sustained patterns of interaction with religious individuals and organizations. Finally, spiritual atheist scientists—those who identify as spiritual but do not believe in God or consider themselves religious—represent the smallest group of atheists at 6 percent of atheist scientists in each country. In the next three chapters, we consider each group in turn, beginning with those *seemingly* most like the New Atheists: the modernists.

3

"I Am Not Like Richard"

Modernist Atheist Scientists

Modernist atheist scientists are the kind of atheists that most
everyday people tend to imagine. Among modernist, culturally
religious, and spiritual atheist scientists, modernist atheists hold
views on religion that most resemble those of the New Atheists;
they do not believe in God and they have no official engagement
with religion or spirituality. Yet, even these scientists vary widely
in what they think about religion—making this group the one with
the most considerable within-group variation.

Let us be extra clear here: even the group of atheist scientists
we thought would be *most* like the New Atheists do not actually resemble them very much. While some modernist atheist
scientists are indeed fierce critics of religion, others do not have
negative views of religion—they think it provides helpful societal symbols and rituals, or they simply give no thought to religious belief. Many also regard New Atheist discourse on the
relationship between science and religion as unproductive and
even damaging to science. Of atheist scientists in the U.S., we
categorize 66 percent as modernist atheists, and in the U.K.,
we categorize 73 percent of atheist scientists as such. In what
follows, we hear first from modernists who embrace views most
like the New Atheists before turning to their more moderate
counterparts.

Varieties of Atheism in Science. Elaine Howard Ecklund and David R. Johnson, Oxford University Press.
© Oxford University Press 2021. DOI: 10.1093/oso/9780197539163.003.0003

Problems with Religion

Modernist atheist scientists do indeed find religion problematic. And a significant minority think that religious scientists are inconsistent at best and unable to do excellent science at worst. Other modernist atheist scientists think that religion is not only problematic for science but problematic for society broadly. For example, modernist atheist scientists are much more likely than other atheist scientists to describe the relationship between science and religion as one of conflict—62 percent in the U.S. and 63 percent in the U.K. By comparison, only one-third of modernist atheist scientists in the U.S. and U.K. see science and religion as independent. Exceedingly few think that science and religion can be used to help support each other.

Religion Is a Problem for Science

Modernist atheist scientists tend to believe that religion and science are inherently incompatible. In the view of many of these scientists, religion and science can't be reconciled because when they try to explain the world through fundamentally opposed methods and worldviews, religion comes up not only wanting, but also sometimes in direct conflict with scientific explanations. For example, a U.K.[1] biologist described this conflict as "irredeemable" and "irresolvable," expressing dismay that religion does not require the basic laws of physics and chemistry to apply all the time. "So, if God can make people walk on water, then what's Price's Theory of Fluid?" he asked. "When I *do* think about religion, then I apply my scientific methodology, the most I can make of it is [that] it is mass psychosis or neurotic behavior by inadequate personalities." Another U.K. biologist[2] similarly expressed the idea that religious belief is irreconcilable with science because the worldview of science is fundamentally different from the worldview of religion, describing the two as "opposite sides of the coin at some level." He expressed his perception of the difference between religion and science, saying, "If you feel that there is a higher power involved, then it seems to me like it gets a little bit messier in terms of trying to understand fundamentals of science."

Other atheist scientists we spoke with described viewing science and faith as fundamentally irreconcilable because religion seems to inculcate a spirit of "just believing things" without proof or evidence, a way of thinking that they saw as antithetical to science and learning science. As a young person, a physicist[3] from the U.K., told us, he was "just trying to . . . accommodate my naïve religious views of the time with what I was learning from science and I saw that actually the two things don't go together very well." It wasn't the outputs of religion and science that he found incompatible, but rather the ways of thinking. "It's about the methodology that in religion you should have faith . . . without any proof of it essentially," he said, "while in science it's the opposite, you know. You shouldn't believe in things without evidence." These modernist atheist scientists are engaging in the ultimate form of *scientism*;[4] they believe that there is no aspect of nature or the physical world that cannot ultimately be explained through science.

Another key concern of modernist atheist scientists is that religious belief is actually harmful to or even impedes scientific advancement. One U.S. graduate student in biology,[5] for example, said:

> I think that religion in general . . . has a very long record of hindering science, the advancement of science. When we've challenged that the Earth revolved around the sun and not the other way around, the church [was] almost violent against that idea and shut it down and that's happened many, many times and I think that history . . . it just leaves me a little bitter.

These gripes about religious perspectives impeding crucial scientific advancement aren't all relegated to the distant past. A U.S. biologist[6] worried about the effect of religious people on the fight against climate change:

> I understand religion as . . . random ideas they have . . . that people . . . kind of adopt over the time . . . they believe many things, and these things really go against the natural evolution of science, and makes things harder, because if you're trying to help to—for instance, to fight against climate change or to fight against the rise of the temperatures or whatever, or any other global change and people are saying that it's . . . not happening, then things are harder.

In other words, this scientist thinks that because religious people may be less likely to believe in crucial crises facing the world, it makes the job of science in fighting those issues more difficult. And because religions come with a preset collection of axioms, certain areas of science that might require one to think outside those strictures might challenge religious people. If the goal of science is to "push the limit," then religious people who "already have a limit" might find it too difficult or provocative.

Religious Scientists Are Inconsistent

Because they thought religion impedes science, it was sometimes hard for modernist atheist scientists to wrap their minds around how religious scientists can embrace both faith-based ways of knowing and scientific ways of knowing. One lecturer in biology,[7] for example, did not think being religious made someone a bad scientist per se, but given his understanding of religion and science, this modernist scientist did not understand religious scientists. He described being religious as living a kind of double life:

When you take an approach, when you ask a question, you make observations, you obtain data and you use that data to make predictions about future events. That's the formation of theory. And if you're given teachings about the world around us from a religious point and you say, this is it, this is all you need to know, don't worry about anything else, it's all contained in this book, you don't need to know anything else to live a good life, I think that's when you have problems. . . . You're living two lives, in my opinion. There's a Jekyll/Hyde about it.

A biologist[8] from the U.K. also thought that it would be hard for religious people to practice science, because science does have definable limits and religion does not, such that when religion enters science, science cannot occur:

> The claims of science, such that they are, work because I do something in a restricted setting, therefore it becomes universal but applicable, that the experiment illustrates in a small context, how everything works, that what happens here and now isn't a special occurrence which isn't repeatable. If you don't have that view, then no science is possible. If something happens as if by magic, as it were, it's a one-off event and it's not repeatable, then no science is possible.

This biologist felt that the empiricism of religion relies upon a belief in "magic," and this belief kills scientific processes. He concluded, "if you don't accept the basic premise that the specific can be extrapolated, why are you bothering doing science?"

Other modernist atheist scientists, however, took a stronger view of their religious scientist colleagues. These scientists thought that certain religious beliefs—such as creationism—are directly in conflict with religious ideas, and this tension can actually have an impact on the work of religious scientists. "You see a complex system and instead of trying to get into to . . . develop the pathways or into maybe the reason why all of these different organisms have the same system," suggested one U.S. biologist,[9] religious scientists might be inclined to "just say, oh, this is very complex and there is no reason for it except that God put it there and so you don't actually strive to understand the world around you. How is that science? That's just not." Another U.S. biology professor[10] told us he thinks religious scientists:

> have to be slightly schizophrenic, and say, "I'll deal with the world on a scientific basis and use data and experiments to determine or tell me how things work, but there's another aspect to the world that I won't require that level of evidence for.

(It's important to point out here that in our broader studies of scientists, both religious and nonreligious, we have never found a religious scientist who used a "God of the gaps" approach as a reason to ignore scientific findings.)[11]

Similarly, a physicist[12] in the United States contrasted the constricting methodological nature of religion with the freedom of scientific investigation:

> Sometimes religion limit[s] your imagination, right? You want to be a really good scientist, you have to be creative, have imagination, have curiosity. However . . . most of the time they put a circle around what you should think. The best of the science is . . . go out of [the] box, right? To find, to push the limit. [With religion], you already have a limit.

Another professor[13] also strongly emphasized the dissonance between a person working in a field that has defined theorems and axioms while holding beliefs in a transcendent deity:

> I cannot understand how you can be religious and a scientist at the same time, because I do think that when you are a scientist

you try to find explanations and so the concept of having something that can never be explained, I don't understand how you can live with that as a scientist.

One physics researcher from the U.K.[14] had a different perspective on the impact of religion on scientific work. Instead of discussing the philosophical differences between the two, she instead recalled the practical implications on daily work in the lab, referencing particular religious groups who have problems with animal models or human embryonic stem cells:

> They certainly have come across in the U.K., you know, certain students who will come in and will refuse to do different types of the experiment. Because they don't want to work on this particular animal, or they don't want to work on stem cells, or things like this. So I think for some people, yes, it can influence exactly the type of science they do. There are some people who won't work on animal experiments at all. Yet they want to be a biologist.

A physicist[15] also pointed out to us that biologists, in particular, might have trouble balancing their religion with their profession. "So I think . . . there could be a problem for biologists, if they—I imagine that if there is, if there is a biologist who believes in creation, this can interfere with his professional research or her research." A U.K. biologist[16] repeated this idea: "I would think somebody who has a deep-seated belief in, for example, creationism, I think is not compatible with being a good cell biologist or a good biologist." She continued, arguing that the nature of biology can come into conflict with one's religious worldview. "So much of that [good biology] is dependent on hav[ing] a real understanding of evolution and the genetics and so on and really I think appreciating evidence-based science is very difficult in that [religious] background."

At the same time, some scientists described physics as presenting the most unique challenges to people with traditional religious backgrounds. One graduate student in physics,[17] for example, told us, "Right, you're not going to see any astronomers talk about how the universe is 6,000 years old. Like that's complete—it's never going to happen, it's 13.6 gigayears. And we might argue about if it's 13.6 or 13.7, but it's 13 point something billion years old." She went on, telling us that facts like this might keep people with very traditional beliefs out of science altogether or certainly make it difficult for those within science who have more traditional beliefs to express them. She explained: "And so with scientists, you're not going to see those types of people because . . . it just doesn't at all match with science." She then took issue with Christians, in particular, who might object that the details of science show evidence of God's action in the universe. "You're not going to see it. So mostly it's just like, we have this pretty nice picture of how the universe has evolved in time, and so when you start actually looking at the nitty-gritty details of it, you're not going to find big God questions in them. You are going to find, 'CO_2 formed at this, or CO formed at this redshift, and this tells us something about the ionization states and how many stars there were.'"

One professor[18] concurred with the opinions just cited, telling us that religion smuggles biases that make science difficult to pursue with intellectual integrity:

I guess my own bias is they might be a little closed off to certain possibilities because they have some preformed possibilities. Whereas as a scientist, I think you need to have a very open mind to look at all the data and interpret it no matter which direction it's going to take you. So, if I had evidence saying that the Earth was a certain age and I'm faced with that evidence, but I have a bias that the Earth cannot be that age then I think the non-religious person would probably consider the possibility that the data is saying what it's saying.

In her mind, the biases and preconceived limits that religious scientists place on the world—in the area of creation, especially—can actively hinder scientific enterprise. In the words of one U.K. lecturer,[19] "If there's a conflict, that must mean at least one of the models must be wrong."

Religion Is Harmful to Society

The public is fairly split on the impact of religion on society in general. A 2010 survey of more than 18,000 individuals in 23 countries found that 52 percent agreed that "religious beliefs promote intolerance, exacerbate ethnic divisions, and impede social progress," while 48 percent agreed "religion provides the common values and ethical foundations that diverse societies need to thrive."[20] A more recent 2019 Pew survey[21] found that 55 percent of people in the U.S. specifically believe that religion and religious institutions do more good than harm in society, 53 percent think that they strengthen morality in society, and 50 percent feel that religion mostly brings people together. In contrast, only 20 percent of people in the United States thought that religion does more harm than good. Unsurprisingly, citizens of secular countries are more likely to see religion as having a negative impact—71 percent in the U.K., for example— while citizens of more religious countries tend to have a more positive impression of religion's role —65 percent in the U.S., for instance.

The idea that religion is regressive and dangerous to society is a central argument of the New Atheists. It is "not an exaggeration to say that if the city of New York were suddenly replaced by a ball of fire, some significant percentage of the American population would see a silver-lining in the subsequent mushroom cloud, as it would suggest to them that the best thing that is ever going to happen was about to happen: the return of Christ," U.S. neuroscientist Sam Harris,[22] one of the most prominent New Atheists,[23] has said. "It should be blindingly obvious that beliefs of this sort will do little to help us create a durable future for ourselves—socially, economically, environmentally, or geopolitically. Imagine the consequences if any significant component of the U.S. government actually believed that the world was about to end and that its ending would be glorious. The fact that nearly half of the American population apparently believes this, purely on the basis of religious dogma, should be considered a moral and intellectual emergency." Steven Weinberg, a U.S. physicist and Nobel prize winner, has said that, "With or without religion, good people can behave well and bad people can do evil; but for good people to do evil—that takes religion,"[24] and he believes, "Anything that we scientists can do to weaken the hold of religion should be done and may in the end be our greatest contribution to civilization." The New Atheists do not see positive influences or benefits of religion. "Many people are good. But they are not good because of religion," Victor Stenger, another American physicist and New Atheist, has said. "They are good despite religion."[25]

Many modernist atheist scientists see religion as broadly damaging to society. In our interviews, we heard from a number who hold views on religion similar to those of the New Atheists, embracing similar rhetoric and illustrating their points with extreme examples that present religious groups and individuals as irrational, prone to authoritarianism, and actively harmful to society through the promotion of damaging or dangerous ideologies. The most prominent theme underlying the concerns about religion we heard from scientists was the idea that religion is harmful because it erodes or undermines rational and critical thinking, particularly the form of cognitive rationality promoted in the sciences. "There does seem to be some conflict between science which says 'Critical thinking is the most important thing, you know ask questions, ask why do we think this, where did this come from, let's figure the whole thing out,' as opposed to a lot of religions seem to dictate, 'This is truth because we say so and you're not allowed to question that,'" a U.S. doctoral student in atomic physics[26] said. One U.S. physics student[27] provided an example of this kind of thinking, recalling conversations that she and her colleagues had about religious people and their views that oppose scientific consensus: "It's usually complaining about basically religious people in other parts of the country doing stupid things. There will certainly be frustration about people who don't believe in climate change, or who think that the Big Bang is wrong. So it's usually sort of all of us thinking those people are idiots, which is not the most charitable thing, but it reflects the frustration."

Sociologists have historically emphasized the importance of cognitive rationality to social order in modern society, brought about largely through research, particularly in science. Early sociologist Talcott Parsons specifically noted that "research is the societal activity which is most 'purely' oriented by cognitive rationality . . . as the activity moves from this core, the value pattern of cognitive rationality has to be integrated in relevant societal subsystems with other values."[28] And W.E.B. Du Bois saw rational thought as a tool to fight against racism, which he characterized as "age-long complexes sunk now largely to unconscious habit and irrational urge."[29] Some scholars believe that a rise of cognitive rationality will eliminate the influence of religion in society and bring about a better society overall.

A U.K. biologist[30] talked about the ways religion destructively affects how people think about the world, giving leaders weapons of intimidation, fear, and indoctrination that keep people from thinking for themselves. He described religion as "a way to control en masse if imposed properly" and discussed the Middle Ages as an example of a time when faith was wielded to exert political and social control. He explained that he sees parallels between this history and modern times, asserting that "religion realizes that science can explain everything that it can, and I think it's obvious that religion, or various religions, are changing their explanation of things in a very poor attempt to rationalize what's going on." He went on to express his fear that this way of thinking could be detrimental to humanity as a whole, saying that he worries about "where are we going to be in, ten, twenty years' time as a species?" Wars, terrorism, and extremism based on religion were always at the forefront of this scientist's mind, and these anxieties motivated him to stay informed about world news and events related to Islam, Christianity, and what he saw as other fringe groups in the U.S. "I feel I need to inform myself on both sides so that I'm actually fearful of the future as a person of the species," he said, "and I do think that one of the major causes for that fear and worry is religion from all sides." He pointed to violence between Protestants and Catholics in Northern Ireland as an example that hit particularly close to home, which he described as "just absurd when you think about it. . . . And it does worry me."

"I think they're just less—their standards for evidence-based decisions may be weakened because an aspect of their lives doesn't require evidence at all," a U.S. astronomy professor[31] told us about the influence of religion on believers. "And so an authority figure, which may be a religious leader, can make a statement that isn't questioned." Another U.K. biologist[32] told us that when religious people are moderate, open-minded, and reasonable, then religion does not harm society, but when religious fundamentalists raise their voices, society can experience division and regression:

> Everything that we do, pretty much, is run by previous science that has turned into technology and medicine and that's why we live so long and we live so comfortably in the developed world. . . . And some of the more vehement religious sort of fundamentalists would like to eliminate large aspects of scientific understanding and perhaps almost go back to a medieval type of lifestyle, which I find a bit disturbing. So, if it's [religion is] moderate and if it's intelligent and conscious, then I think it's very beneficial. If it's irrational and hysterical and dogmatic then I think it can be very dangerous and can block science.

Another U.K. scientist[33] bemoaned the division that religion causes, arguing that religious people insert themselves in conversations in which they have no authority or expertise and further polarize society around those issues:

> I think sometimes it's certainly in some, some of the high profile cases you see in the court systems, both in the U.S. and here, you end up with a system where quite frankly a lot of people who really shouldn't stick their oar in feel the need to do so because they have a religious viewpoint, they feel the need to point out, come down on one side.

The scientist continued, providing as an example the Terri Schiavo[34] case (a highly publicized right-to-die lawsuit in which, after entering a persistent vegetative state, Schiavo's parents and husband fought a prolonged legal and public battle over whether or not Schiavo's feeding tube and life support should be removed):

> You have a situation where realistically the only people who need to be involved are the partner and perhaps children and parents and yet it became this huge thing that sort of people coming out on one side and the other of debate, various religious groups said "we're going to lend our voice to either this side or that of the argument" and it, it sort of became less about her in the end and really she was the only person that should have mattered. . . . I feel lots of times it should be about the people who are involved in the personal circumstances and be involved with, and it, it should just contain much more empathy than the whole thing does. Which becomes very much about "well my book says this" and "your books say that."

While anecdotal evidence and historical cases certainly lend some support for such critiques, the most rigorous data offer exceedingly narrow evidence in support of the idea that religion undermines critical thinking broadly. For example, data from the General Social Survey—a nationally representative survey of U.S. citizens conducted each year since the early 1970s—show that no religious group differs from the nonreligious in terms of their propensity to seek out scientific information or their knowledge of scientific facts.

Appreciation for the Positive Societal Role of Religion

Though many modernist atheist scientists we spoke with were critical of religious thinking and expressed concern for its impact, others acknowledged a positive societal role for religion. Lindsey, an atheist biologist we met in the U.K.,[35] argued that science and religion mostly fall into separate categories, yet she is respectful of religion and can see why people may be drawn to it. "I can understand why people go to religion. Because I think it's just a way . . . to console themselves," she said. "And to give greater life satisfaction." Throughout our discussion with her, she noted many ways she thinks religion is beneficial for its adherents. In particular, she recognizes value in religious institutions and rituals. Religion, she told us, creates a "nice environment" for gathering and leisure:

> Along with the idea of religion, there's the practice of religion, right? For getting together, the baking, the tea parties, the little summer fests that they have here, and we go to the little summer fests . . . as well, because it's a place to meet people, and for the kids to play, and it's a nice environment. . . . I'm never going to set foot in church or anything like that. [But] it's part of culture here, and it's part of—a long, pervasive culture, and I can understand the cultural aspects of it.

"It's a set of rituals around which many of the key stages in [the] life of people in certain parts of British society are framed," another

U.K. biologist[36] told us, explaining his view that religion could provide societal cohesion through its practice. "So I do enjoy, I do enjoy those. I mean I get something out of it," he said.

A biologist from the U.S.,[37] who worked previously in Europe and Asia, also expressed that there is value in religious practice and religious knowledge, saying, "Even if you just try to learn the Bible, learn all the stories, it's not a bad thing. And sometimes it's curious when you haven't seen the other side. Another group of people and what they believe, how are things going. . . . And you learn something actually, I think the Bible has told many good stories, that you try to be a good people, and that's pretty good." This scientist saw value in the lessons that the Bible teaches, "even if you don't believe" in the underlying source of the moral teachings, she explained.

And a graduate student in physics[38] likewise discussed the value she found in studying religion in school: "I generally find it quite interesting because it's quite well-rounded. . . . So you learn about a lot of them [religions] and so you do see there are a lot of connections between all of them and then there's an overarching [feeling of] just wanting to be for something else that you can't explain. Yes, it's quite— yes, I do think that religious education was a good thing in schools."

We showed earlier that a significant subset of modernist atheists view religion as a harmful force undermining the status of critical thinking in society, but another group of modernist atheists perceive religion as playing a positive and valuable societal role —acting as a force for good, encouraging people to act more charitably, for example[39]—even as they personally reject religious faith. Religion "can motivate people into doing charitable work. . . . [Y]ou see people [doing] good things, helping the homeless, things like this," another biologist in the U.K.[40] said. "Very often it is religious people who are motivated to do that. . . . [N]et, I don't think it [religion] is either good or bad." One physicist[41] even told us that she has "always also had a respect for religion" because of the way that religion taught her acceptance and tolerance.

Other modernist atheist scientists point to the meaning they think religion can confer to believers, and some even seem to convey a sort of envy for the individual fulfillment and sense of belonging they see religion as offering. "I would like to join something like this that didn't have that particular meaning to it," one scientist said of church. "I mean, there's basically like sports from church where you can go and like hang out with a lot of people and all be excited about the same thing and then go home back to your families like. And I'm not into either" he laughed as he referenced sports. But, "so I wish there was like a third thing that was for like, you know, the people who aren't into sports or church [laughs]. . . . I wouldn't mind something like that, but this author, Alain de Button, is like putting together this like atheist church, which I've sort of been curious about, but haven't gone and checked it out at all. But I do . . . like the idea of a community and feeling connected to a community." Another modernist atheist biologist[42] discussed how her mother's experience gave her a more positive view of religion: "Yeah, she believes; she definitely believes in God and . . . like she's the one who has made me much more tolerant of believers than I was when I was twenty." She continued, telling us that religion gave her mother the ability to work through the death of her sister: "[My mom's] sister died and it was horrible for her and the thing that gets her through it is the idea that she will see her sister again one day in the afterlife. And that's really important to her and I think she's totally wrong, but I have no way of knowing that for sure. She might be right and so . . . it really helped her get through it. . . . I think I relaxed a lot on my kind of bulldog attitude towards—how can people be so like brainwashed by—by this obviously false, you know, premise."

Other modernist atheists view religion in generally positive terms but believe that religious extremists make religious movements look bad. One U.S. physicist[43] had otherwise neutral views of the influence of religion on science, with one caveat: "Yeah, so I would think people who are very, very extreme in their religious beliefs. So I'm thinking of very, very far right-wing Christians in this country or very conservative Muslims in other parts of the world where their religious belief is so much more a part of their life than a lot of other people in the world practice their religion." He concluded, "I would think that their worldview is so shaped by their religion that they would not be able to accept things that don't mesh with that religion, and so, you can't be a scientist." One U.S. biologist[44] continued this sentiment:

In general, it's positive but the problem is the extreme people are usually the loudest and people only take notice of when things are actually having a big effect— . . . the [few big extremists] helped to villainize the bigger group.

In the U.S. especially, religious extremists have an outsized influence on the perception of religion. Our research has shown that when nonreligious scientists in the U.S. think about "religion," they tend to think "evangelical"—or, more specifically, a biblical literalist who embraces creationism.[45] It is possible that nonreligious scientists, who likely have little direct contact with religious individuals or experience with various religious groups and institutions, fall back on the stereotypes of religious groups and extreme voices they most often hear in public discourse and culture.

"I Am Not Like Richard Dawkins"

British New Atheist Richard Dawkins has written a lot about religious belief and religious extremism and his view that religion fosters division, idealism, fervor, and thus destruction, both physical and moral. He writes in his book *The God Delusion*,[46] for example:

Our Western politicians . . . characterize terrorists as motivated by pure 'evil'. But they are not motivated by evil. However misguided we may think them, they are motivated, like the Christian murderers of abortion doctors, by what they perceive to be righteousness, faithfully pursuing what their religion tells them. . . . The take-home message is that we should blame religion itself, not religious extremism—as though that were some kind of terrible perversion of real, decent religion.

Dawkins also stated, in response[47] to a 2017 terrorist attack in Manchester, U.K., that "religious faith really does motivate people to do terrible things. If you really, really believe that your god wants you to be a martyr and to blow people up, then you will do it. And you will think you're doing it for righteous reasons. You will think you are a good person." Dawkins is often intentionally acerbic and confrontational in his descriptions of religion: "Do you really mean to tell me the only reason you try to be good is to gain God's approval and reward, or to avoid his disapproval and punishment? That's not morality, that's just sucking up, apple-polishing, looking over your shoulder at the great surveillance camera in the sky, or the still small wiretap inside your head, monitoring your every move, even your every base thought."[48] He elaborated in another part of *The God Delusion*, "The God of the Old Testament is arguably the most unpleasant character in all fiction: jealous and proud of it; a petty, unjust, unforgiving controlfreak; a vindictive, bloodthirsty ethnic cleanser; a misogynistic, homophobic, racist, infanticidal, genocidal, filicidal, pestilential, megalomaniacal, sadomasochistic, capriciously malevolent bully."[49] He summarized his thoughts on religion with the pithy, "One of the truly bad effects of religion is that it teaches that it is a virtue to be satisfied with not understanding."[50]

"There have been many books I've been influenced by one way or the other, even when I was religious," one U.S. modernist atheist biologist[51] told us when we asked about influences on her view of the relationship between science and religion. "I have to say that a book that—it didn't make it, but it crystallized my atheism—was *The God Delusion* by Richard Dawkins."

While many U.S. scientists have been influenced by Richard Dawkins' ideas about religion and its relationship with science, Dawkins has had an outsized impact among U.K. scientists. Many of the modernist atheist scientists we interviewed there mentioned him by name. Some of these scientists agree with his argument that science and religion are inherently in conflict and champion his ideas and combative approach. Others value Dawkins' role as a provocateur, even if they don't agree with all of his ideas. In their view, having such a provocative public persona allows him to agitate—and engage—an audience beyond the scientific community, which allows modernist atheist ideas to have a far greater impact. As these scientists see it, his aggressive mode of public engagement gets him attention, permitting him to promote scientific and critical thinking in a highly visible manner—and thus, they see his approach as positive and important. Many modernist atheist scientists appreciate that Dawkins is invoking the cultural authority of science in public dialogue. In their view, he is a torchbearer for cognitive rationality. "He's probably a bit more reactionary than I would be, but I think he has quite an important place in society actually in prodding things," said a biology professor[52] from the U.K. "I wouldn't say that [his books are] always right, but they're certainly an interesting take," another lecturer in biology[53] explained. "It sparked some interesting conversations actually . . . which probably wouldn't have happened otherwise . . . it was thought provoking definitely, which I think is always a good thing."

Another modernist atheist, a graduate student in biology[54] in the U.K., said he had "very much followed . . . Richard Dawkins and other debates," and told us:

> That isn't necessarily to say that I agree with [his] beliefs or maybe that's it, because I believe that maybe Richard Dawkins and these

kinds of people who are actively rocking the boat, ... they may be talking in ways which are bigger than science can answer, but that isn't to say that I [don't] find it fascinating.

Yet, we also found modernist atheist scientists who critiqued Dawkins' approach. His engagement style is rejected on the grounds that, as we have written elsewhere, it "promotes the scientist over science, derision over diplomacy, and ideological extremism over dialogue."[55] One cluster of U.K. scientists we interviewed said they are not fans of the way Dawkins talks about religion *or* science—feeling that, in arguing for the superiority of science over religion, his public engagement misrepresents what the process or method of science can do and the questions it can answer. A number of scientists we spoke with believe Dawkins does not properly convey the limits and limitations of scientific inquiry. One professor of biology[56] told us:

> He's a fundamental atheist. He feels compelled to take the evidence way beyond that which other scientists would regard as possible. ... I want [students] to develop [science] in their own lives. And I think it's necessary to understand what science *does* address directly.

Another U.K. biologist[57] described Dawkins as a fundamentalist:

> There are, of course, fundamentalists on both sides of the fence. We've got Richard Dawkins—[do] you know Richard Dawkins?—on the atheist side, who says we should talk about atheist rights and we should promote ourselves.

A postdoctoral fellow in biology[58] elaborated on this point, saying, "I think you have to be very careful about stripping away people's beliefs without offering anything in return. And I think a lot of the sort of most vociferous promulgators of religion—and I was in Oxford when Richard Dawkins was there. He's probably the most guilty of it." A third biologist[59] further criticized Dawkins, saying, "I've read a lot of Richard Dawkins. I find his approach to critiquing religion a bit annoying. He's really critiquing the concept of a creator I suppose more than religion. . . . [But] I don't think he's careful enough with his words to make that clear."

While a major segment of U.K. scientists are atheists, and the majority of these atheists are modernists, many of them do not feel that Dawkins' approach to religion and atheism reflects the majority of the scientific community. They expressed concern that Dawkins oversteps the bounds of science and makes it seem that scientists are dogmatic rather than open. Importantly, many also believe that he gives the public the false impression that all scientists do—or should—share his kind of atheism.

We can see that modernist atheist scientists are not monolithic—they are, in fact, not all like Richard Dawkins and the collection of writers who call themselves the New Atheists, but instead have varied and often nuanced perspectives on religion. Some do clearly think religion is problematic, pointing to its negative effects on science and society as justification for their antipathy. Others have a more moderate view, acknowledging the neutral character of religion and blaming only extremists for the conflict between science and religion. Still others think of religion as a net *good* in society—a moderating and moralizing force that brings people together. This, however, does not paint the full picture of atheist scientists in the U.S. and U.K. We found that many such scientists go a step further than their modernist counterparts that like religion—they actually engage with religion, attending services, participating in church life, and even sending their children to religious schools. It is to this category of atheists that we now turn.

4

Ties That Bind

Culturally Religious Atheists

Focus on the Family, a conservative Christian organization that promotes public policies informed by a biblical worldview, has a "Q&A" page on its website that addresses the question: "Can an atheist and a believer build a strong, lasting marriage?" In short, their response is: no. "Because in the final analysis," the group argues, "the challenge you're facing is bigger than a mere difference of 'religious opinion' . . . it's a matter of dramatically contrasting *worldviews*. And when worldviews collide, the results can be devastating for a marital relationship."[1] On Reddit's atheism subreddit—what some characterize as *the* online public square for atheists—a post entitled "Atheist married to Christian" describes an atheist man, married to a Christian woman, who is earnestly seeking advice on how to respond to marital conflict related to his lack of belief.[2] One response, representative of many others, reads in part: "Unfortunately if you ask me your marriage was already over quite a long time ago."[3]

Atheists and religious individuals appear to share similar views on faith and marriage (perhaps to the surprise of each group!). As one participant in sociologist Jesse Smith's study of atheist identity development observed, "I couldn't date a true believer . . . because I couldn't respect how they approached the universe . . . [this] line of thinking is not something that I would want in my intimate life."[4] For many atheists and believers, a consensus has emerged that a person's faith, or lack thereof, is a "dealbreaker" in a relationship. In their view, relationships between atheists and religious believers

Varieties of Atheism in Science. Elaine Howard Ecklund and David R. Johnson, Oxford University Press.
© Oxford University Press 2021. DOI: 10.1093/oso/9780197539163.003.0004

will inevitably succumb to a clash of worldviews, and it is difficult to fathom what type of partnership could even be possible absent a shared standard of moral or spiritual values.

Religious Atheists?: Atheist Scientists Engaging with Religion

Most atheist scientists tend to be segregated from religious individuals in their day-to-day lives.[5] The tendency of atheists to primarily interact with other atheists, or religious individuals to primarily interact with other religious individuals, reflects the adage "birds of a feather flock together." Sociologists refer to this as homophily,[6] a pattern in which individuals with similar character- istics tend to interact primarily with one another. Homophily can be driven by an internal preference—such as a religious individual who prefers to socialize with other religious individuals. It can also be driven by opportunity. For example, atheist scientists may pri- marily socialize or collaborate with other atheist scientists simply because most of the people around them at work are also atheists. Our data does not allow us to determine whether the relative lack of interaction between atheists and religious individuals is driven by preferences or opportunity structure. For modernist atheists who are anti-religious, we can assume preference is the major constraint to interaction. For modernist atheists who are neutral or not anti- religious, opportunity structure may be the larger reason for lack of interaction with religious individuals.

When we surveyed atheist scientists, we found a second category of atheists—atheists whose lives are characterized by sustained and even intimate interaction with religious individuals and organiza- tions. Ethan is an atheist and a professor of biology[7] in the U.K. who initially seemed to embrace a New Atheist view of religion. He told us he agrees with Richard Dawkins, "that religion is a bit of a silly thing and it's not based on any evidence." (In his book on evolution,

Ethan thanked Dawkins for inspiring his interest in science.) Like many atheist scientists, the media attention given to religious groups opposing the theory and teaching of evolution annoys Ethan, and he is often involved in public debates on the topic. "I did actually get quite cross about these people making these silly comments and all this nonsense," he told us. "I felt that I had to make a response—you know [I was] kind of an angry young man."

Given this background, we were surprised to discover that Ethan sends his children to a religious school in the U.K. where they even attend church services as part of their schooling. Like many of the atheist scientists we have already heard from, Ethan is respectful of religion, admits there are things he likes about religion, and acknowledges positive aspects of religion, even though he does not adhere to a religious tradition or have personal faith:

> You can see that there are benefits of going to a church and having a youth club and all those kinds of things, so I don't really see it's my view to go and force people to change their beliefs in the same way as if they follow a particular football team or whatever. You know . . . in some ways you could say it's a matter of taste.

Why might a self-described atheist, one who admires Dawkins and debates religious people, send his children to a religious school or admit to religion benefitting children? Is he inconsistent, hypocritical, or acting on some understudied logic? And is he alone? In this chapter, we examine conversations we had with scientists like Ethan, who have sustained patterns of interaction with religious individuals and organizations. We call this group of atheist scientists "culturally religious" atheists.[8] These are not atheists simply on visits to Notre Dame or touring Buddhist temples on vacation. They are atheists who, despite their irreligion, exhibit various social ties to religious individuals, either through marriage, formal affiliations, attendance of religious services, religious schooling of their children, or—at times—through their work, and these ties make this group of atheists *the most unlike* the New Atheists.

For example, when we asked an atheist biologist in the U.K.[9] if religion currently has any place in his life, he said he considered religion to be like a social activity for him. "I am a sidesman [usher in the Anglican Church] in the local church," he told us, "and I enjoy going, and I love reading the lesson, and I enjoy the liturgical aspects. I enjoy the opportunity to engage in meditations or contemplations, slow down, think about what you're doing, those kinds of aspects." Further, imagine an atheist undergraduate who is no longer religious, but when she returns home to the religious networks of her youth and dinner is on the table, the prayer soon begins: "The eyes of all look to you, oh Lord, and you give them their food in due season. You open your hand; you satisfy the desire of every living thing, Amen." While as an atheist she fundamentally rejects the beliefs behind this ritual prayer, she nevertheless closes her eyes, takes the hands of her family around her, recites with them in unison, and enjoys the ritual. This simple example represents a broader pattern among atheists that we refer to as cultural participation in religion. To some, this cultural participation might seem contradictory and confusing. Though some atheist scientists do not self-identify as religious or hold religious beliefs, they nonetheless participate in religious traditions, communities, and organizations for seemingly nonreligious reasons.

When we surveyed atheist scientists in the U.S. and U.K., we found far greater associations with religion than we expected. In the U.S., 29 percent of atheist scientists can be considered culturally religious atheists, meaning they have sustained interaction with religion through participation, prayer, or marriage. One in seven are formally affiliated with a traditional religious denomination, and more than 16 percent attend church on occasion. Nearly 9 percent of U.S. atheist scientists pray from time to time. In the U.K., we find that 21 percent of atheist scientists are culturally religious. One in ten are formally affiliated with a traditional religious faith. About

12 percent of U.K. atheist scientists attend church on occasion, and 3 percent report praying on occasion.

Here we take a closer look at culturally religious atheist scientists. First, we hear from atheist scientists who are still involved in religious traditions, particularly those who were raised in a religious faith, to learn why, even though they no longer embrace the beliefs of their youth, they continue to participate in faith rituals and services. Second, we examine atheist scientists who culturally immerse themselves in religious networks for reasons related to their background, social status, or other affiliations. Finally, we look at atheist scientists who are married to someone religious.

(Non)Traditional Pathways

As we showed in Chapter 2, a sizable majority of atheist scientists were not affiliated with a religious tradition during their early life stages. About 55 percent of atheist scientists in the U.S. and U.K. did not belong to a religious tradition at age 16. The typical atheist in our sample either grew up without religion or abandoned the faith and traditions to which they were exposed through their families. But a significant minority of atheist scientists in the U.S. and U.K. forge less conventional pathways over the years of their lives, continuing to identify with and belong to a religious tradition, but without belief in a higher power. Traditionally, researchers have not paid much attention to those who no longer believe in a higher power but continue to identify with a religious tradition.[10] For many years, for example, U.S. researchers would not even ask questions about religious participation to those who checked the box "atheist" on a survey. In the U.S., most individuals grow up religious and stay that way in all senses of the term—belief, affiliation, and practice. In the U.K., there is some debate over whether both religious believing and belonging are in decline or whether individuals are engaging in what sociologist Grace Davies calls "believing without belonging."

Either way, atheist scientists who are formally affiliated with a religious tradition do not fit neatly into the more conventional religious or secular pathways of either country.

Affiliating with a religious tradition is one of the core dimensions of cultural participation. Our data shows that 15 percent of atheist scientists in the U.S. and 10 percent of atheist scientists in the U.K. affiliate themselves with a religious tradition even though they also self-identify as atheists.[11] When we examine the religious traditions with which cultural atheists currently identify, we see that the major religious traditions include Catholicism, Judaism, and Hinduism (each at 17 percent), followed by Buddhism (16 percent) and Protestantism (10 percent). Fewer than 1 percent of culturally religious atheists identify with Islam or Orthodox Christianity.[12]

A minority of culturally religious atheists we met identify with a religious tradition other than that of their youth. An atheist professor of astrophysics,[13] for example, told us he grew up Catholic, but left his childhood faith tradition and now attends Quaker services and even practices some elements of the religion. He explained that he is not drawn to practice as a Quaker due to their beliefs, but instead because he feels that in the Quaker tradition, he has found a group of people that he can relate to and feel comfortable around. In other words, he is attracted to, and prioritizes, the sense of belonging he finds in the religious tradition—and it is this group solidarity that motivates his religious practice. As he described:

> It's not necessarily an embrace of Quaker tradition. I think it's more like that I feel at home. The Quakers are not a church, they're not a religion, right . . . it's more like a haven where people—like-minded people, liberal, spiritual people—would go . . . and I feel that's a nice community there, that they're all people that have similar spiritual or religious opinions.

Our interviews with culturally religious atheists revealed engagement to be chiefly driven by a desire for connection and group belonging. What does it mean to belong without believing? This sometimes translates to participation in practices that others might see as contradictory or antithetical to atheism, such as religious rituals or even prayer.

For atheist scientists who were raised in a religious tradition, continued religious practice after they have suspended religious belief is sometimes motivated by a powerful desire to maintain group solidarity with the members of that tradition, and the cultural heritage of the religious tradition remains salient absent belief. In the U.S. and U.K., atheist scientists from a number of faiths exhibit this approach to their religion. A professor of physics in the U.K.,[14] for example, discussed the lasting impact of her childhood faith. She said she no longer considers herself a practicing Catholic and does not actively attend church, but still practices certain Catholic tenets—focusing specifically on the concepts of sacrifice and diminishing the importance of the self—and admitted that she can still see the remnants of Catholic teachings in her outlook and actions, even in her professional life. She views some of these remnants as beneficial and is indifferent to others, saying, "I think, probably, having a respect for hard work has benefitted me professionally." Although she is no longer a practicing Catholic, "I continue to respect [the people who practice]," she said, "and the religion is part of the culture. But that's to do with the people . . . when you respect people, you respect what they care about and what they do. And if what they care about is saying their prayers, then you respect that."

Jewish Atheists—a Case Study

Cultural identification with a religious tradition in the absence of belief is somewhat unconventional in the U.S. and U.K., but not without precedent, especially within Judaism and Catholicism. Writing about the Jewish community in the U.S., sociologists

Steven Cohen and Arnold Eisen argue that American Jews "choose what to observe and what not to observe; they also decide, and take it for granted that they have the right to decide, with no one able to tell them any decision is wrong, when to observe, how to observe, and how much to observe."[15] As sociologists of religion have long noted, choice is central to most of American religion. And thus it is unsurprising that secular Jews—and scientists in particular—choose tradition but not belief.

This approach to religion, based on heritage and group identity, is prevalent among atheist scientists who were raised in Catholic, Hindu, Buddhist, and Jewish traditions, who are all likely to belong without believing.[16] Here we focus on atheist Jewish scientists as a case study in how religious heritage and group solidarity[17] are the key factors that drive such attachment. Group solidarity is especially important for groups of people who have experienced genocide and continue to experience intense global persecution.[18] A number of Jewish atheists, for example, report being culturally religious, or what we would qualify as religious in terms of practice and group identity, but not in terms of doctrinal adherence. They often attribute their continuing affiliation to their desire to maintain a connection with the broader global Jewish community.

"Practice is important. I guess for me, that's part of spirituality. [It's] very non-supernatural spirituality," a British physicist,[19] who identifies as Jewish, said of his current religious participation. This participation, he said, "reminds me of the history of a particular group of people, that I have some connection to. . . . It's partially a kind of old prideful thing, that I am connected to this lineage of people going back 7,000 years."

This sometimes means participating in practices that others might see as contradictory or antithetical to atheism, such as religious rituals or even prayer. Jewish atheists often attribute this seeming discrepancy to the desire to maintain a connection with the broader global Jewish community. One U.S. graduate student in physics[20] told us:

I'm an atheist and a Jew, and that works for me, but it certainly wouldn't work for a lot of Jews *[laughs]*. But I think that there are some group of Jews who believe that it's fairly compatible to follow these other prescriptions and still be an atheist. . . . I am culturally involved in religion, so I run the [Jewish student association], I have a decent group of friends here who are all Jewish and who I met as part of this policy of trying to have a life outside of lab *[laughs]*. And I celebrate Jewish holidays, I celebrate Shabbat some weeks, my boyfriend is Jewish . . . and that's a significant part of our relationship.

"It's very important, I kind of almost sometimes wish it were more important," another atheist physicist in the U.K.[21] said of his relationship to Jewish practice. "When I forget to fast on Yom Kippur, I feel a little guilty—and not because God will strike me down." He grew up in New York City, he said, where, in his words, half the children he knew had a bar or bat mitzvah. In his home, he told us, the belief aspect of religion was almost never discussed, but religious practice was very much a part of his family's daily life. "I think it's possible to be in some sense spiritual because those stories are interesting, without actually believing in any of the supernatural stuff," he said. Today, he thinks of himself as ethnically but not religiously Jewish. "If I go to a wedding or bar mitzvah . . . of somebody, you know, then I'm happy to at least do my best to attempt to read along with the Hebrew, as I once was able to do."

A graduate student in biology[22] felt similarly. She said, "I do a bunch of high holidays. I am going to do Passover, but it is totally cultural. I like being a part of that community and I don't think that if I didn't do it God would be angry." She does not view this absence of belief as lessening the importance of the practices, but instead ascribes an alternate significance to them, one related to shared heritage, cultural identity, and a sense of belonging.

An American professor of physics[23] told us that because Judaism played such a central role in his early life, he cannot fathom *not* being Jewish. "Jewish people tend to value education and reading and learning. And so that aspect of the culture, which is not directly part of the religion, but maybe comes out around because of it, has had a huge effect on me," he said. He does not take part in many Jewish practices or traditions—and when he does, he does not ascribe a spiritual or religious meaning to them—but he feels strongly about the importance of self-identifying as Jewish and draws a connection to group identity and solidarity, saying:

> I'm Jewish by declaration, basically. And by culture. I find myself responding in ways that are part of how I grew up. I grew up in a largely Jewish neighborhood. . . . My friends were mostly Jewish, I know some Yiddish words. I feel sympathy with the way Jews experience the world, and I decided some years ago that if there was any question about whether I was a Jew or not, then I was a Jew *[laughs]*. And if you don't like it, stick it.

"I certainly don't have any practices; I usually don't even know when Passover is. Or Rosh Hashanah," said another atheist scientist[24] who told us he was Jewish "by culture really." Despite the fact that he rarely engages with religious belief or practice, he finds other ways to stay involved in his faith. He clarified, "I'm part of a group that in a way is making Jews more visible in the world. In this group, on Friday nights they would do a Shabbat service. And . . . they said, 'Anybody who'd like to do this who's never lit the candles?' And I had never lit the candles, and so I did it. And it was quite moving for me actually. Yeah, I think it is part of reconnecting with my father and my culture I grew up in."

Does participating in this sort of religious ritual, and others like it, make an atheist religious? What does it mean when an atheist covers his eyes and says, "Blessed are You, Lord our God, Ruler of the Universe, who has sanctified us with commandments, and commanded us to light Shabbat candles?" Is it religiously significant when an atheist is emotionally moved by a religious ritual or prayer? These questions are not exclusive to Jewish atheists. It is not unusual, for example, for atheist scientists at Oxford and Cambridge universities to attend Evensong, an Anglican service of evening prayers and hymns. One U.K. biologist[25] explained:

> Since I joined this college [at my university]—I would call this more spirituality than religion—but I attend occasionally the Sunday Evensong in the chapel here and that is a very kind of pleasant place for musical reflection. You can take it to any level. I go there as a sort of, I'd say perhaps a spirituality-type experience where there's beautiful choral singing, candles, and atmosphere.

Another atheist biologist[26] told us:

> I enjoy going to church for the suspension of disbelief, for the theatrical experience, for reading, for the liturgy, for the magnificent stories and the mythic quality of those stories, which is intensely spiritual. I mean, that's a real experience. But what I don't get, what I've found myself just unable to reach, was the kind of communion with the deity.

These atheists recount their experiences at church, despite their unbelief, as profoundly affecting and beautiful. They not only willingly engage in religious rituals, but gain spiritual fulfillment from them, while still identifying as atheists. Perhaps this emotional connection can further help explain why some atheists maintain a connection to religion. Sociologist Randall Collins[27] studied connection and feeling and theorizes that emotional energy is the basis of all rational action. As he explains, "By recognizing emotional solidarity with a group as the primary good in social interaction, it is possible to see all such value-oriented behaviors as rationally motivated toward optimizing this good." In other words, for some atheists, participation in religious rituals makes sense as a piece of their cultural heritage, as a way to feel emotionally connected to a group, and thus build and solidify group identity, membership, and belonging.

For Cultural Capital

Some atheist scientists are motivated to participate in religious organizations and practices as a way to cultivate and signify cultural capital—what sociologist Pierre Bourdieu[28] saw as the soft social assets, such as aesthetic tastes, respect for formal culture, appropriate use of language, and manners, that signify membership in status groups.[29] Any competency could be a form of cultural capital. While knowledge of religion, like knowledge of literature or painting, might not be economically useful, many people associate these forms of cultural capital with higher social standing.

Through our interviews with atheist scientists, we found that the decision some make to enroll their children in religious schools often appears driven by factors related to status, culture, and class. Ethan,[30] the British professor of biology we introduced earlier, for example, said he sends his children to a religious school where:

They had to go to church services. They had to pray in the assembly and all these kind of things. That kind of ground against us a bit, but . . . they're getting otherwise a good education and the amount of irritation that we get out of this is not dreadful.

Sociologists such as Bourdieu[31] and Annette Lareau[32] argue that privileged parents (which would include individuals in prestigious lines of work like science), have greater agency to utilize their own cultural capital to secure their children's social position. Some atheists may see placing their children in religious schools as allowing their children to cultivate their own cultural capital. When we talk with atheist scientists who send their children to religious schools, it seems they do so in part because of the social advantages they feel their children will get as a result.[33]

One of the key ideas from theories of cultural capital is that individuals who possess high levels of cultural capital profit more from education and have greater success in the pursuit of career paths in prestigious occupations. Many atheist parents choose to send their children to a religious school because they believe institutions aligned with religion can help their children acquire competences that will open up opportunities for occupational mobility.

Scientists, for example, routinely engage in public speaking and thus place a premium on this competency for social mobility. For Ethan,[34] the cultivation of this skill is part of his rationale for sending his children to a religious school:

> They used to have [. . .] Christmas services and the children would participate in them and I thought, you know, . . . they're actually getting confidence to speak in public and those kinds of things, and so it was kind of useful.

Ethan also likes that religious education allows his children to develop knowledge of the classics and society's cultural heritage. "Children . . . should know a little bit more about the Bible and about religion because most of Western culture can't be really understood unless you have a bit of an idea about the Bible," Ethan[35] explained. "In the same ways you can't if you don't know a little bit about Greek mythology and Greek civilizations."

Ethan was not the only atheist we interviewed who sent his children to a religious school; rather, the theme appeared repeatedly in our study. Most send their children to religious schools for the same reasons Ethan does: educational and cultural enrichment. One biologist[36] we spoke with went so far as to chair the board of governors for his daughter's religious school:

My daughter, she's thirteen. Again, it's not—she knows what my views are. I've never had a deep conversation about it, because I don't want to drive her in any particular direction. She needs to make her own mind up. She was never christened or anything like that. She goes to school, which is—has a Christian ethic to it. I'm chair of governors of a primary, which is a CFE controlled primary school, which I always thought was quite ironic, but it's a CFE controlled primary school and they have an atheist as a chair of governors booster, but still—but I did it for education purposes.

Just because atheists might send their children to religious school, however, doesn't mean that they want them to absorb the religious aspects of their education. The astrophysicist from the U.K.,[37] whom we met earlier, who was raised Catholic, described his children's reactions to their education in Catholic schools:

My children, so I've got three boys and they all went to Catholic church school, because they're—the better schools when we came here was the Catholic church—and they are, these—these schools are very traditional Irish conservative Catholicism and the boys immediately [laughs] reacted by saying this is rubbish, right, at an early age, so they always developed it [atheism] much faster than I, you know, and so they are—but they still, you know, have the same, I think, ethics.

British sociologist Grace Davie[38] argues that even as religiosity has declined in the U.K., religious schools remain popular due to "vicarious religion," in which religious schools assume responsibility for religious socialization. The theory is that parents who send their children to these schools view religious institutions favorably not because the parents embrace religious belief but because of the contributions they believe religion has made to the public good, national culture, and heritage.

Strange Bedfellows

Many atheist scientists are not opposed to dating, partnering with, or marrying someone who is religious. About 15 percent of atheist scientists in the U.S. and 11 percent of atheist scientists in the U.K. have a religious spouse. From our interviews with atheist scientists, we learned that differences in faith and belief are often not ignored in these relationships. Rather, these relationships often lead atheist scientists to attend religious services or embrace raising children with religion in the home. For example, we spoke with a professor of physics in the U.K.[39] who grew up as a practicing Christian, but now identifies as an atheist. He told us his wife is a practicing Catholic. When we asked, "What does a marriage comprised of an atheist and Catholic look like?" he told us that he attended "church very regularly, particularly when [the children] were younger, more so [actually] than some supposed Catholics who didn't bother going." He added that he still attends with his wife "from time to time" now that his children have left home. His attendance is not tied to his own solidarity with a religious tradition, as we have seen in other cases, but as a way to support the values of his family members.

Another British professor of physics[40] explained that marrying a practicing Christian led him to attend church with his family often but has not changed his lack of belief; he remains an atheist. "I guess, to be honest, I am kind of an interested observer," he said. "So I think my wife would call herself a Christian; I wouldn't . . . I enjoy what I hear there, and I find much of it very helpful, but I couldn't—couldn't honestly call myself a Christian." Another atheist scientist[41] who recently got engaged to a Catholic man described her plans to participate in the Rite of Christian Initiation necessary to convert to Catholicism, saying, "I'm willing to be a Catholic, but really I'm still probably not a Catholic considering I don't really have any of those type of beliefs, I'm just an atheist sitting in a Catholic church." These atheist scientists, and many others who have a religious spouse, are not only tolerant of their partner's beliefs, but are also willing to participate in their partner's religious practices, approaching their faith with respect and negotiation.

For other atheist scientists who are married to a religious partner, religious practice in the household is more of a shared cultural practice. When both spouses practice religion in a cultural way, religion can be easier to navigate than in relationships in which one spouse is an atheist and the other is traditionally religious. In some cases, religion can become an avenue of exploration through which they can discuss different ideas or observe different practices. Take, for example, the case of an atheist graduate student in biology[42] in the U.S. who identifies as ethnically Jewish but not religiously Jewish. He is married to a Japanese woman who identifies as a Buddhist but does not consider herself a practicing member of her religion. "We do Jewish holidays at my uncle's house, so I mean yeah, we participate in that stuff, but our day-to-day life is not influenced at all by this," the biology student told us. "It's just occasionally we will . . . participate in religious things and it's the same when we go back to Japan and visit her family." Belonging to different religions does not cause any strife in the family or marriage, he said. Instead, their differences lead to discussions in which each can explore the other's beliefs and practices. He recounted some of their past conversations, saying:

> It's philosophical. Because there are certain things about Judaism that she doesn't understand and some things she will never understand, and sometimes she says, "Why do you guys do that?" . . . And the same thing goes for me, like, "Why does this happen?" And so we have those discussions—kind of philosophical discussions, and I'll explain to her what's going on, definitely

ceremony, and the same for her. So it's really just—it's real exchange that's kind of ongoing, but she's interested in this stuff. She doesn't believe it, but she thinks it's interesting, and the same goes for me; I think whatever the stuff they do back home is interesting, too.

For many atheist scientists, different from the warnings of Focus on the Family and Reddit, being married to a religious spouse does not lead to animosity or strife. Things can become more complicated, however, when children are brought into the equation. How do atheist scientists and their religious partners decide if their children will be raised with religion? For New Atheists, the answer is simple. In an article titled "Don't Force Your Religious Opinions on Your Children,"[43] Richard Dawkins writes that, "it is high-handed and presumptuous to tie a metaphorical label around a tiny child's neck . . . at the very least it negates the ideal, held dear by all decent educationists, that children should be taught to think for themselves." In another article, New Atheist writer James Di Fiore[44] states, "If I ever found out an adult made my child go through any of these psychologically damaging experiences, my response, ironically, would be biblical."

Yet the culturally religious atheist scientists we spoke with expressed a different view. Many do not hold strong convictions about raising their children without religion and said they have decided to raise their children in the church or to expose them to religion. "In our case, we, I think, came to agreements fairly early on. And that was I would support her in bringing up the children in the religion," a British professor of physics[45] we met earlier told us of making a decision with his Catholic wife. "It's inevitable, of course, that the kids at some point—and they're now, they've reached that point—are old enough . . . to make up their own minds," he said. "And so, you know, in [a] sense that was also a part of the deal . . . that at some point then they would make up their minds . . . and that the pressure on them wouldn't be too intense, if you like, when that point came."

Several other atheist scientists we spoke with described similar situations, in which they agreed to let their children be brought up in their partner's religion, and many also described this decision about religious upbringing as conditional, continuing only until the children are old enough to make their own decisions about their religious practice and beliefs.

The professor of theoretical astrophysics[46] and participating Quaker we met earlier said he and his wife raised their three children as practicing Quakers, but did not enforce their practice once the children were old enough to think for themselves and express their own opinions. While his youngest child still attends the Quaker church, his two older children have decided to stop attending. One is now a practicing Muslim who implores his father to read the Quran, while the other has decided to no longer attend church altogether.

Atheist scientists who agree to raise their children with religion seem less concerned with pushing their nonreligious or atheist beliefs on their children than they are with allowing for critical thinking and self-decision on matters related to religion and faith. The culturally religious atheist scientists we spoke with do not seem to care whether or not their children ultimately end up religious or atheists. They merely want their children to be given the option to decide for themselves what—or if—they believe.

5

Spiritual Atheist Scientists

On a freezing Friday morning in London we set out to interview Jacob, a microbiologist.[1] Despite the cold weather, the walk to the university was delightful. Parents were dropping their children off at a local daycare, the children seemingly oblivious to the cold, as children often are.

Jacob studies how life could develop from inert molecules. When we met him, he was dressed in a vintage shirt, black skinny jeans, and red Converse, looking like he would feel just as comfortable at a local pub as he would in a lab. He was relaxed, funny, and personable, punctuating our time together with humor and wit, yet he quickly became reserved and insightful as we began to talk about the connections between religion and his work as a scientist. "I would define myself as a spiritual atheist," he told us. "So I'm pretty happy in saying that I'm an atheist, although there's a huge sort of conversation that we have anyway with anyone who would be interested in engaging me in what that means!"

Jacob's claim made us pause. He had already told us he was not religious and had frequently praised Dawkins throughout our discussion. We asked him to explain a little bit more about his views of atheism and spirituality. We were especially curious about how Jacob saw spirituality in relation to the transcendent, since many think that spirituality necessitates a belief in a god of some type. According to Jacob:

> Consciousness is not readily observable, but I believe in the fact that we are conscious. So there are plenty of phenomena like that that are intangible and not readily observable—that can't be

Varieties of Atheism in Science. Elaine Howard Ecklund and David R. Johnson, Oxford University Press.
© Oxford University Press 2021. DOI: 10.1093/oso/9780197539163.003.0005

explained in terms of atoms and molecules, at the level of their atoms and molecules, but they . . . cannot act off the atoms and molecules. There's nothing supernatural about them. They're sort of—there are intangible driving forces to the universe above and beyond elementary forces like gravity and electrical attraction and sort of emergent properties of syntheses and holistic phenomena that are real, but we can't reduce them to their atoms and molecules.

When Jacob spoke of spirituality, he spoke of intangible realities that imbue wonder, motivate his work, and are beyond observation. At first blush, his spirituality made him seem like an outlier among the atheists we had encountered in our fieldwork. But as we traveled across the U.S. and U.K., we discovered many more atheist scientists for whom spirituality is important, often as it relates to their research. We call these scientists "spiritual atheists."

Atheist scientists define spirituality in ways that vary both from each other and from the general U.S. and U.K populations as a whole. For spiritual atheist scientists in the United Kingdom and the United States, spirituality means constructing an alternative value system without affiliating with a specific religious tradition. Spiritual atheist scientists also see themselves as different from other atheist scientists who have no spirituality (modernist atheists) as well as atheist scientists who participate in religion (the culturally religious atheist scientists we introduced in the last chapter).

In both the U.S. and U.K., spiritual atheist scientists often had different conceptions of spirituality, its use, and where to find it. Many spiritual atheist scientists *found* spirituality in their scientific work, frequently describing a sense of awe that they could only explain in spiritual terms. Others *used* spirituality as a tool to navigate life, death, and personal meaning. Some had a kind of personal spirituality that they discussed in contrast to prevailing notions of spirituality outside the scientific community. In this chapter, we examine what atheist scientists think it means to be spiritual and what, for them, spirituality looks like in practice. We begin by considering what we know from research on spirituality in the general public and past research on spiritual scientists.

How Everyday People Construct "Spiritual but Not Religious"

To understand spiritual atheist scientists, we must understand spiritual but not religious (SBNR) individuals more broadly. In some sense, spiritual atheists follow a worldwide trend in which more and more people are rejecting the guidance of a particular belief system or religion to instead stake out their own spiritual territory.[2] According to a recent survey, 27 percent of adults in the U.S. and 11 percent of adults in Western Europe say that they think of themselves as spiritual but not religious.[3] Sociologists have been studying new forms of spirituality in the general population for several decades. Some versions of spirituality exist among established networks of organized religion, while others exist largely apart from traditional denominations and practices.[4]

While scholars have used a continuum of definitions for the term *spiritual*, they have identified some general themes that characterize spirituality in the general public. For example, both those in the U.S. and the U.K. seem to link spirituality to interaction with some form of theism.[5] Americans, in particular, also seem to pick and choose among religious traditions to develop their sense of spirituality, which scholars have called a socio-syncretic spirituality. For example, the same individual might do yoga, engage in prayer, attend a Shabbat dinner, and follow a form of Buddhist meditation during the week.[6] However, their counterparts in the U.K. are much less likely to do this, partially because, in spite of the presence of the Anglican Church, when compared to the U.S., those in the U.K. are less religious across the board. To put this in perspective, when active spiritual practices are tallied up, U.S. SBNRs may engage with theistic religious practices just as much as, or more than, those in the U.K. who actually identify as Christians.[7]

These "spiritual entrepreneurs" complicate the narrative that nations secularize as they become more industrialized because people who are spiritual but not religious are most often found in the most developed nations.[8] They usually reject traditional conceptualizations of the sacred, instead inventing their own definitions, uses, and practices of the spiritual and transcendent. Some scholars have argued that, beginning with the baby-boomer generation in the U.S. (those born between 1946 and 1964), many people were not as willing to submit to an organized religious authority.[9] Instead, individuals began to value the freedom to choose and combine different cultural resources, either religious or secular, after self-reflection.[10]

Most people who are SBNR in the general population link spirituality to connection with a transcendent force or God.[11] For example, a professor of historical theology, Linda Mercadante, writes in her book on modern SBNR people that "few, if any" of the people she interviewed "considered themselves outright atheists" and yet she goes on to say that "I could not assume that when they spoke of the sacred, a transcendent reality, a divine dimension, or even used the word 'God' that they meant a 'personal' god, an 'Almighty' who created the world, hears prayers, takes an active part in earthly affairs . . . even if an eternal non-bodily Presence—is nevertheless someone with whom they can 'have a relationship.'"[12] Some use meditation, prayer, or even activities like belly dancing to stimulate their spirituality.[13] Others tie their spirituality to more belief-oriented tenets, like trusting in guardian angels or believing in the existence of the afterlife.[14] The key feature is that new spirituality is fundamentally individual and shaped by each practitioner. While it might be tempting to view spirituality as simply a watered-down religion that has benefit only to the practitioner, many scholars see it as providing a way to connect with the transcendent *without* the confinement of organized religion and *with* the possibility of generating concern for others. All of these perspectives point to spirituality as an important development in contemporary religious life.[15]

Spirituality among Scientists

And yet, we have wondered, how does this play out among scientists, those who seem the most committed in societies to reason, consistency, and the material world as the only world, those least susceptible to the ethereal? When we turned to the lives of scientists, we found that in contrast to the popular view that most scientists are avowed atheists, and that all others are religious,[16] many U.S. and U.K scientists reject the idea of a binary choice between total rejection of the transcendent and total acceptance of the divine. For example, Elaine found in a study of U.S. scientists at top research universities that, like the general public, spiritual scientists describe their spirituality in many different ways. Some see very little inconsistency between their work as scientists and their personal spirituality, thinking of the latter as an extension of their work and a motivating factor for improving the lot of humanity.[17]

The nascent literature on scientists' spirituality suggests that scientists might perceive being SBNR as an innovative way to even navigate tensions between science and religion. Scientists who are spiritual but not religious prize consistency between their work as scientists and their spirituality. This may be different from everyday people then, who seem to almost prize using disparate logics and practices to create their own sense of spirituality.

And U.S. scientists who are spiritual seem to be different from most other spiritual Americans in that their spirituality does not always have a connection to a particular belief in God or even to the transcendent at all[18]—and since spirituality does not *require* belief in God in the same way these scientists perceive religion does, many scientists see spirituality as actually more congruent with science than religion is. For example, spiritual atheists in both the U.S. and U.K. are the least likely among the three categories of atheism (modernist, culturally religious, and spiritual) to embrace the conflict view of religion and science, and a greater proportion agree that some dialogue can exist between science and religion, though there are dramatic disparities between the two countries on this view.[19]

How Spiritual Atheist Scientists Construct Spirituality

In our survey, when we asked atheist scientists in the U.S. and U.K. not only what they believe about God, but also whether they consider themselves to be a spiritual person, we found that about 6 percent in both countries consider themselves spiritual. However, in the earlier survey Elaine conducted just among U.S. scientists at top research universities, when she asked atheist scientists whether they were *interested* in spirituality, more than 22 percent said they were.[20] Spiritual atheist scientists told us that they do not believe in God because they believe the existence of God cannot be definitively proved through empirical scientific evidence, and they have found ways to understand and experience spirituality without God.

For example, when one biologist[21] was asked how she understands the terms *religion* and *spirituality*, she explained:

I guess religion implies that one believes in some kind of God or something like that, and it's often an organized group that sets up writings and moral values and rules. . . . I don't belong to any religion now. I always assume that people who have spirituality believe in God and they think of it that way. Personally, I believe in nature, and I get my spirituality . . . from being in nature, but

I don't really believe there's a God, so I don't consider it's necessary for what I do or how I behave.

While our data clearly indicate that spirituality is mainly an individual pursuit for academic scientists—an activity that helps them understand themselves better, sometimes in relation to wider humanity—for some, it is not individualistic in the classic sense of making them more focused on themselves, but rather something that motivates them to more deeply understand and positively have an impact on humanity. "I think [spirituality] is all of nature, including human being[s] and all the population, . . . coming together to form a force nobody can stop," one physicist[22] from the U.S. told us about her notion of spirituality. "That's what I believe, and also I believe in doing good things."

Awe and Wonder in Science

Across the narratives we heard from spiritual atheist scientists, we see two primary frameworks for evaluating the world in spiritual terms. One framework casts spirituality in emotional terms of awe and wonder resulting from a scientific understanding of the world. The other defines spirituality in relation to the limits of scientific knowledge.

Some atheist scientists described a spiritual impulse marked by compatibility with the scientific method and a visceral sense of reverence experienced through science. Their scientific work—exploring and examining the mysteries of the universe and the complexity and vastness of nature in the quest to understand them—lead these scientists to experience awe and wonder that both flows from their work and is wholly beyond it.[23] Jacob,[24] for example, told us that his work made him admire the grandeur of the universe:

Spirituality is an aspect of that. So . . . understanding the world, the universe, in a way that is more about acceptance . . . whether it involves an explanation or not of why—why the universe is and why we are in it. [*eight-second pause*] I think—this is somewhat subjective—but there's some sort of sense of wonder.

For Jacob,[25] this sense of grandeur from which his spirituality flows is inseparable from his professional identity as a scientist. He told us further:

I think my definition of spirituality is fairly nebulous. I mean, I'm not doing my work to cure disease or do something that will fan a startup company. . . . I suppose the goal is—the goal of my work is service to humanity—to impart a sense of wonder, and so that's sort of all part of my worldview that encompasses spirituality as well.

Jacob thinks that this feeling of commitment to something larger than the self, a sense of awe, should motivate the pursuit of science, and he looks to recruit students who want to join him in pursuing science out of a sense of awe.

One U.K. physicist[26] described the overwhelming feelings he experiences when he looks at pictures of deep space and contemplates the vastness of the universe and nature:

I think some of the photography that you see from Hubble . . . it is all that wonder in a way, but I mean it is absolutely mindboggling the scale, the . . . clusters and super-clusters, galaxies, and knowing what that means. I mean that, if you made me think about it, that's awe. I mean it was awe to see the birth of a child, *your child*, and . . . I'm a sort of [an] outdoor person so sometimes I get that when you're outside. I go [what I call] "Imagineering." So you're on the top of a mountain and you get a sense of all that . . . it gives you an emotional response and it makes you think how amazing the world is and . . . that is a sort of spirituality.

A biologist[27] from the U.K. similarly tied spirituality to both the scientific enterprise and the feelings that can result from experiencing the grandeur and magnitude of nature. She told us:

I also think spirituality is something you feel. You can sometimes instinctively feel things that are bigger than you. Like when you're standing by the seaside or when you see some big place, beautiful nature place. When you feel small and you see the world, you know. Or even when you see something really amazing as a scientist, you figure out some amazing mechanism and you suddenly understand it, that can be like a spiritual moment. So I can't really give you a definition. I kind of have a vague feeling. I know when it happens, but I've never had to define it.

What Science Cannot Explain

Some of the spiritual atheist scientists we interviewed also felt that to be spiritual was to come to terms with *the things that science cannot explain*. These atheist scientists used spirituality to deal with the limits of science, articulating the dividing line between what is knowable and what is not as the boundary between science and spirituality. For example, the physicist[28] in the U.K. who described the wonder he feels looking at pictures of space also told us:

I think in some ways I almost do feel spiritual about things, which is strange. I think our understanding—our scientific understanding—is [limited], even though we think we know an awful lot. To me it does feel that there are things that we don't understand . . . which might come under the title, "Spiritual." So I've seen situations where you almost think there is a link between people or animals or there's something else that our current sort of worldview, the whole depth of physics, chemistry, biology, doesn't actually explain. So I wonder whether there is some extra component that we really haven't got yet and maybe that might be something which other people see as metaphysics.

He went on to muse about his definition of spirituality and its relationship to the limits of science. In particular, he thought human objectivity was very limited:

I suppose it depends what you define as spiritual. . . . I do have a sort of spiritual element. I think . . . there's only so far that you can, if you're a complete human being that you can, be objective about the world. There's this feeling about yourself and your existence and on that level, neurology, physics, it doesn't remotely explain the sense of being, and to me . . . that is what I would call spiritualism and what I would say science can't explain.

A biologist[29] we talked to, also in the U.K., similarly spoke about coming to spirituality through her scientific work. In her case, the complexity and intricacy of living organisms led her to believe there is something "more" out there than science can explain. She said:

I have to teach about how a genus is switched on in the cell and it's so complicated, and there's so many proteins in the world, we don't even understand all of it, and this is happening thousands of times literally every hour in your body. . . . And so then I think, wow, you know, how can this just have happened through random chance? So, yeah, it does make me wonder if it's some kind of more—some force, you know, that is responsible for life. I think that . . . you can look at everything at many different levels, and I just don't believe that we are capable of seeing the big picture at all. You know, we don't have enough time in our lives, we don't see at all the different levels. I do believe in something bigger than us as it were which creates order in the world. I guess I do believe in that, but I mean I couldn't define it to you.

A Consequential Spiritual Atheism

The salience of spirituality varies among the spiritual atheist scientists we met during our fieldwork. Some see elements of the world in spiritual terms that can involve emotional responses, but they neither embrace specific practices they would define as spiritual nor view their spirituality as consequential to other behaviors. For other atheist scientists, however, spirituality manifests through day-to-day practices and relationships.

Spirituality Found through Scientific Work

One of the central ways that spirituality is embodied in practice for atheist scientists is seeing spirituality as a day-to-day component of their scientific work. This is part of what is considered an identity-consistent spirituality, a relatively self-consistent set of beliefs and practices that fits well with previously existing identities, like one's identity as a scientist.[30] This form of spirituality may flow from holding a job that constitutes an all-encompassing identity, or what others have called an identity with master status.[31] Because of the all-consuming nature of science as a profession, scientists may be moved to integrate their spirituality with the already coherent sense of self that is organized around their work as scientists. For scientists, then, spirituality cannot be compartmentalized, since their work overlaps with their identity to such a large degree.

For example, during our interviews with spiritual atheist scientists, many of them talked about seeing their work exploring the mysteries of the natural world or the cosmos as spiritual in some way. For them, spirituality is less a tool to use to better oneself, and more of an inescapable facet of their jobs. "The spirituality part of it . . . is why you do the work," one physicist[32] told us. "I'm interested in understanding how the universe began, possibly what its long-term future is going to be. I think those are certainly spiritual questions." Some scientists suggested that because they do not believe in God, they are able to feel a special sense of reverence for the natural world. They believe that leaving God out of the equation frees them up to admire the complexity of the natural world, contemplate it with awe, and praise it. "I think there might be some respect for life that biologists have because it's so amazing . . . we see the inner workings and it is amazing," one biologist[33] told us. "I don't call it religion though." Another biologist who specializes in ecology[34] told us about his exaltation of nature and how it has made him a better scientist:

> Because I believe that nature is the mother of everything, and we belong to nature. I think we are extremely connected to nature and I try to focus my job, my work, on getting [to know] Mother Nature. Not only protect it but also kind of restore it or bring it back to better conditions that we had before. So I feel a really kind of totally personal connection, a spiritual connection with nature. That kind of motivates a lot of my work.

We also met atheist scientists who described how their spirituality not only motivates their work as scientists but is also expressed through their work.[35] According to one U.K. scientist,[36] for example:

> When one has a sense of spirituality, one believes that humanity has kind of a common consciousness. I don't mean that we read each other's minds or anything like that, but something that links us together is the fact that we are conscious animals. And we share dreams and hopes and some feeling of being human, which for me is part of a human spirituality. And, therefore, one would

like to contribute to the well-being of that general humanity. And one way to do that is to work in a profession [like science] whereby you're trying to help other people. . . . I would find it hard to reconcile my day-to-day activity if I didn't feel like I was doing something that would benefit the greater human good.

A number of spiritual atheist scientists emphasized how the spiritual impulse could be compatible with the practice of science. A graduate student in physics[37] said:

I think it's okay to believe in religion and also do science because in some sense you're basically worshiping your physics. *[Laughs]* If you believe that God started everything and created the initial state of the universe and all the physical laws and set it in motion, you're basically saying that God is the physics. At that point you're still looking for God, but you're also looking for science and it's the same thing.

Spirituality as Moral Engagement for Scientists

Both the U.S. and U.K. are secular in the sense that state and church are separate,[38] yet in both countries religion is seen as the foundation of morality, politics, civic participation, and even citizenship.[39] In the U.S., however, there is (generally) a positive perception of religion and (generally) a pejorative stance against atheism,[40] while in the U.K. there is a positive perception of religion and a nondiscriminatory stance toward atheists.

Living in such contexts, where religion is still socially important and where it is linked to ethics and morality in the national cultural imagination, atheist scientists wonder whether religious members of the public will perceive them as lacking cultural and ethical values. As a result, spiritual atheist scientists emphasize the ways in which their beliefs, even without a religious affiliation, inform a set of sound ethical values. While this does not require a codified set of transcendent values, this impulse can lead them to spiritual practices that feed back into the work they do as scientists, whether research or teaching. It can also be a way to distinguish themselves from other scientists who they see as overly focused on their research and scientific pursuits at the expense of helping others or considering others' needs above their own. This form of spirituality is a way for them to mark their scientific work as about more than making money or achieving personal success. One U.S. atheist biologist,[41] for example, told us:

> I don't know if it qualifies as religion or spirituality influencing my science practice but I run a *very kind* lab. That's an emphasis in my lab, an explicit emphasis in the lab. This is how we treat each other, this is how we communicate with each other and with other scientists, and that I suppose is based on my idea of grace. So I've never been explicit about that, even with myself I don't think, until this moment, but I suppose in some ways . . . I model a community of my own creation after some concept of a community that doesn't come from nowhere.

For a number of spiritual atheist scientists, their spirituality undergirds a conviction that they have an obligation to make the world and their communities better places, and they live that out practically in their day-to-day life. In addition to embracing this obligation through their work, spiritual atheists tend to also focus on their relationships with others as a way to practice these values.

When we asked one biologist[42] from the U.K. to define spirituality, she leapt to a description of the importance of investing in people and treating others well. In her words:

> I tend to think you only live once and the bit that makes life worth living is, is not the day to day and the practicality, but it's how

much you enjoy life and what you put into it in your relationships with other people and I suppose spirituality comes into that for me, that there's an essence in people and that's what you're interacting with.

Another scientist in the U.K.[43] talked about how she thinks spirituality has informed the morals she has imparted to her children, teaching them to think beyond themselves and how they should care for and treat other people. She told us:

> I think yes, the way I bring up my kids . . . is without us going to church, but still . . . it's quite spiritual in a sense. The discussions we had . . . the moral code we give to them, there's a lot of spirituality regardless of the fact that we don't believe. . . . They always think about others, and they also think about the things that other people believe, so many other faiths, so I think they are respectful of other people. So I think there is an aura of, you know, of respect or moral kind of conventions that I think is good, regardless if they, you know, we never go to church. So it's the proof that you can be, yeah, spiritual, say, and regardless of the fact that . . . we haven't brought up really a particular religion.

Personal Improvement

Many atheist scientists told us that they use what they consider to be spirituality in their lives to clear their minds, feel better about the world around them, and feel happier and more peaceful in general. For example, one biologist[44] told us that, "Spirituality to me at the moment means yoga and long walks in the woods. So thoughtful time. Time when I have space and time . . . without interruption, without clutter."

Other scientists linked their approach to spirituality very loosely with what they see as Buddhist ideas, particularly meditation.[45]

According to our survey, 22 percent of atheist scientists surveyed in the U.S. and 9 percent of atheist scientists in the U.K. said that they regularly meditate. To be sure, meditation need not entail spirituality and some scientists who engage in it may do so to achieve calm, clarity, or some other physical and cognitive state without transcendent meaning. In interviews, however, we encountered many atheist scientists who did view it as a spiritual practice. For example, a biology professor in the U.K.[46] told us:

> I do meditate from time to time. It's hard to fit it in. In my 20s I meditated quite a lot. . . . And so I got into it a bit then, and I read a certain amount about what spirituality is, what enlightenment is supposed to be. These kinds of things. And I'm still very interested in that. The only ritual, if you like, that I would do is occasional meditation maybe a couple of times a week. And I also consider my running, which I do once or twice a week, as a slightly meditative thing.

Sometimes, the spiritual atheist scientists we interviewed exhibited what sociologists of religion have called a "fuzzy spirituality," a deeply subjective spirituality that pulls from multiple sources, both traditionally secular, like sports, and traditionally religious. For our spiritual atheist scientists though, this does not involve a complex search for the transcendent, but can include spiritual transcendence that is found in nontraditional ways. For example, one scientist[47] connected the discipline and practice that comes from his underwater hockey team to spirituality. He told us about one such experience and the way it made him feel:

> I play underwater hockey. . . . It is not a particularly high-level sport, but, you know, it's [part of my spirituality]. To me, when you're up there and you've got an international match . . . we go through our warm-up routine, we all do it together. We didn't drink for three months . . . before the tournament and we all did

it. And we don't do it because it makes your training better. . . . You do it because it's a sacrifice that then will help you when you're struggling in the water because you know all of your mates are there; they're all ready, they've all done the same thing. Now that's quite spiritual. And so that sort of thing, yeah, and I can see how that works and so I can see why people do that. It makes me understand, if you like, why people have Lent or whatever it is that they do, Ramadan and so on. . . . But from a spiritual point of view, well, I think that's quite spiritual, you know, in some ways.

Spirituality as a Tool for Coming to Terms with Death

A small but growing group of scientists finds that spirituality helps them navigate life, death, and meaning. We found that some scientists do not believe in God but do use language similar to that of religious people to describe some aspects of life. Several atheist scientists told us that spirituality has helped them see and deal with death in another way or has helped them think about their own mortality. According to one such scientist:[48]

I feel a little more comfortable with certain Eastern ideas about individuality as an illusion. When I think about death, for example, I think about it's the end of me, but whatever me was before it was me is going to go back to that. I like this parable of the water going over the waterfall and all these droplets pop out of the water and as they're going down the waterfall they're individual droplets; when they get to the bottom, they go back to the stream. And so these kinds of ideas give me comfort when I think about mortality, but they're not really ideas about a god or anything. But they are ideas about before and after and meaning of life as it is being lived now and that sort of thing.

Similarly, a biology professor in the U.K.[49] told us:

> One of the biggest problems we have in life is death. And I think one of the main aims of spirituality, if you can put it that way, is to come to terms with dying. Because if we can't reconcile the fact that we are going to die with what we're here for anyway, and what it's all about, then it's going to be tough, I think. And so I'd like to spend more time working out my own feelings and thoughts and so on about what that's all about. To me, that's spirituality.

Some spiritual atheist scientists combine their spirituality and their scientific knowledge to derive comfort and reassurance about their place in the universe. In a sense, science becomes the basis of spiritual ideas that help them deal with death. For example, a biologist[50] in the U.S. shared with us:

> So if you think about what happens when you die, I think one of the things that's been most meaningful to me as a science discovery is that all the bits and pieces of us, carbon, atoms, etc., were once out in the stars . . . so if you do calculations as to kind of how you've been built, where those elements come from and where they're going, it's a much more cosmic picture than I might have imagined. And that's been particularly helpful for me when I think about, for example, the death of my grandmother who was not religious, and thinking about "when will we ever be together again?" and I think, well, we will be together again. At some point the carbon in me will [integrate] and we'll be together again!

As we begin to see, atheist scientists create meaning and purpose in and for their lives without religion, a topic we will return to in depth in Chapter 7. But first we will look more deeply at how atheist scientists understand science itself.

6

What Atheist Scientists Think about Science

> Religion is based on dogma and belief, whereas science
> is based on doubt and questioning. In religion, faith is a
> virtue. In science, faith is a vice.[1]
>
> — Jerry Coyne, American biologist

"Don't we already know what scientists think about science?" one of Dave's collaborators, a scientist, asked. This is not an uncommon opinion. We hear this often in our studies of science and religion—that while scientists may differ in their ideas about religion, they surely all have the same ideas about science. Scientists themselves seem to agree that they all share a universally agreed upon definition of "science." But is that really the case?

Over the course of our research, we have often asked scientists what they think about *religion*, but only in our most recent study of scientists did we also ask them what they think about *science*. What we found surprised us: while many atheist scientists have an intuitive understanding of the word "science," they find defining science difficult. "I'm not sure I could express that actually," said one biology professor[2] when asked to define science. "I'm always subsumed in it," stated another biology professor, "so I can't see from the outside, 'what is science?'"[3]

In this chapter, we examine atheist scientists' rhetoric of science with particular interest in two types of boundaries: how they

Varieties of Atheism in Science. Elaine Howard Ecklund and David R. Johnson, Oxford University Press.
© Oxford University Press 2021. DOI: 10.1093/oso/9780197539163.003.0006

demarcate science from religion and whether they place limits on what science can explain.[4] In other words, how do atheist scientists distinguish science and religion as ways of knowing? If science is the only legitimate way of understanding the world, the one and only path to knowledge and truth, are there limits to what science can explain? Do atheist scientists treat the scientific method as sacred, similar to how religious people treat their holy books?

Members of the public and scientists themselves seem to agree that scientists share a universally agreed upon definition of "science." One of the first sociologists to study science and scientists was Robert K. Merton, who theorized the existence of the institutional social norms of science—that is, social mores that govern science that have little to do with the scientific method.[5] He posited that these norms regularized how scientists interact with the public, perform research, and frame their findings. Merton's work has angered some scientists because he seemed to be saying that science is less objective and less consistent in its methods and ways of knowing than scientists themselves like to claim.[6]

Two of Merton's norms are most relevant to the intersection of science and religion. One norm concerns disinterestedness among scientists, as most scientists think it's prudent to act indifferent about the outcomes of their work. This mental detachment involves trying to curb both personal bias and institutional control over one's research. (Religion, some scientists think, stresses devotion, personal investment, and submission to higher authorities, directly contrary to the norm of detachment.) Merton also identified a norm of "organized skepticism" within science, which calls for the suspension of belief or judgment until empirical evidence is fully present.[7] This again helps explain scientists' suspicion of religion, which, many scientists think, requires belief and faith in the absence of evidence, again contrary to science.

Invisible norms, such as those that govern scientific spaces, suddenly reveal themselves when they are broken. The celebrated sociologist Erving Goffman performed many experiments in public places, such as buses and elevators, parsing out the subtle, delicate social norms that govern the taken-for-granted way we behave in these places. In a crowded elevator, for example, it is a breach of a social norm to stand facing the opposite direction as other riders; usually, everyone faces the same direction.[8] However, Goffman pointed out that this phenomenon isn't something that one usually thinks about until they find themselves facing the wrong direction, feeling social discomfort and observing the uneasy glances from other passengers. Facing the *wrong* way in an elevator suddenly brings to our attention that there is a *right* way to face.

Religion in relationship to science may operate in a similar way for atheist scientists. Because religion appears to breach the norms of science with its emphasis on faith in the unseen, lack of falsifiability, and rejection of a purely materialistic worldview, the boundaries around science can become stricter and more apparent when science and religion intersect. Bringing any religious aspect into science—or even the idea of a religious scientist—can suddenly bring to the forefront the supposed *right* way to do science and the right way to *be* a scientist.

For atheist scientists, rejection of religion as a way of understanding the world is central to their identity. As atheists who exclusively rely on scientific knowledge to make sense of the world, the absence of religion is a fundamental element of self-identification. It is this rejection of religion that in part helps us understand what atheist scientists think about science. What science *is,* then, is partially determined by defining what science is *not.*

The Social Construction of Science

Sociologists see science, in part, as a human-shaped cultural tradition, valued for itself and for the effects it can have on people's lives, with norms passed from person to person over time.[9] An institutional logic stressing the transformation of how we understand the world around us governs the scientific community, rather than just scientific facts. One sees this emphasis in the advice a senior faculty member offered to a newly appointed assistant professor of chemistry at an elite university in the U.S.: "All we want you to do is change how we think about science."[10] In some ways, this is similar to a field like art or literature, where innovation and impact on the field are major goals. However, scientific discoveries are governed by explicit criteria, which means that they usually appear objective.[11] In comparison, humanistic fields often seem to cultural observers much more subjective and amorphous than science, because these fields lack explicit criteria regarding what is innovation and what is deterioration.[12]

The work that scientists do often does not just stay within the scientific community. The discoveries, claims, and expert opinions of scientists make their way into the public sphere—where in some cases they are contested by nonscientists. Sociologist of science Thomas Gieryn[13] goes so far as to claim:

> It is in these mediating representations of what science is or what scientists do that sociologists will find a robust explanation for the predominance of science these days in settling questions about the real.

In other words, science and scientists have special authority when it comes to defining and explaining nature and determining what is true and what is false, with the phrases "scientific studies show" or "scientific facts" lending credibility to everyday parlance. (For example, we see this even in toothpaste commercials, when phrases like, "seven out of ten dentists think that this toothpaste is the right one" are seemingly scientific enough to help sales!) And while this special authority is often celebrated and invoked in social institutions outside of science (such as within economic markets like the toothpaste industry), groups in other cultural spaces (such as politics and religion) seek to place limits on scientific authority.

In this respect, what science "is" can partially be understood through the "boundary-work" of scientists.

According to Gieryn, boundary-work involves scientists making an active effort to distinguish *who they are* in contrast to nonscientists, and *what they do* in contrast to what they see as less legitimate methods of inquiry.[14] Science gains cultural authority through boundary-work—a process in which certain qualities, for example objectivity and methodology, are attributed to scientific processes and to those who perform them. This process allows the general public to recognize that science is worthy of intellectual weight. "We learn about science by seeing what is far from it, or near," Gieryn remarks.[15] One of Gieryn's categories of boundary-work is protection of autonomy. This type of boundary-work occurs when the scientific community protects itself and its knowledge from outside powers seeking to exploit scientific authority in a way that compromises the autonomy of science. For example, scientific autonomy is often brought into question when corporate entities try to fund certain types of science, thereby making scientists question whether the funders will have undue influence on the process of scientific discovery or threaten the objectivity of their results.[16] By making sure that science stays unbiased and objective, the boundaries between what is "science" and what is "not science" are strengthened; as a result, science becomes more legitimate and the nonscience that appears to challenge it (like certain religious doctrines) can be dismissed as less reliable.

Definitions of Science

What Do Atheist Scientists Mean by "Science"?

We found three overarching narratives in how atheist scientists define science: one emphasizing science in terms of a strict and defined methodology; another emphasizing the transformation of knowledge; and a third describing science in relation to religion. (Of course, we did not expect atheist scientists to universally bring up religion in their definitions of science—even as it was apparent to them during our interview that the relationship between science and religion was a central concern.)

Empirically Verifiable Observations About the Natural World

The scientific method—that is, a definitive series of steps through which knowledge is acquired—is generally organized around tests, observations, and experiments. The scientific method is perhaps best understood as a series of principles, given that the series of steps that scientists actually employ vary within and across scientific fields. Of the 54 atheist U.K. and U.S. scientists we asked about definitions of science, 24 mentioned the scientific method in their responses. Of these, half (12) identified as modernist atheists, while seven identified as spiritual atheists and five as culturally religious. One scientist from the U.K.[17] defined science in this way:

> So when I say I do science, I say that I follow the scientific method and it's my job to do science; my job is following this scientific method and that may be, by definition . . . to . . . answer problems *[laughs]* . . . so that's starting off with a testable hypothesis, testing the hypothesis, interpreting the results, and developing new testable hypotheses.

One physics lecturer in the U.K.[18] also mentioned the importance of testability, tying it to the ability to falsify assertions:

> It's always difficult to define "what's a good scientific practice?" But there are a few attributes . . . —like, for instance, one important and sacramental preparing, it has to be—you can falsify and

you come up with a system explaining the world around you, which can be falsified. In other words, it can be proven wrong, so if you can't, then you know that you're rich in some kind of irrational belief.

Consider another scientist who defined science by its procedures and principles. What he finds important about science is the way in which questions are asked, which he narrates as being essential to science. He emphasized keeping an open mind and not bringing in other kinds of knowledge that might have an impact on the outcome of an experiment:[19]

> We try not to influence what the answer is even though . . . when we ask the question, we have an idea of what we think the answer might be. We still do it with an open-minded [approach] and that we might be proved wrong. I guess that's going back to basics. Theoretically we start with a hypothesis and we try to prove or disprove that hypothesis.

Defining science by its methods and procedures was one way we saw atheist scientists implicitly demarcate science from religion. Even without explicitly referencing religion, the way they describe science sets it inherently at odds with religion. Science is seen as catalyzing questioning and relies on proof and testing, whereas religion halts questioning and relies on dogma.

Science Is Changeable

Atheist scientists also seemed to view science as having the ability to evolve as more evidence comes into play.[20] Those who share this view believe that science investigates what is empirical and adjusts to new evidence, and many of them emphasized that the field is always searching for new information, even if that information contradicts existing theories and ideas. One graduate student in biology[21] put it like this:

> There are thousands upon thousands of limits. I actually immediately scold students. I have made holes through papers students have sent me, because I've crossed out the word "proved." Any time I see definite, absolutes—it's often in their language—I would just ring them for about five minutes about how that's counterintuitive to what we're doing.

In his view, science provides only more or less evidence in favor of or against a hypothesis or scientific claim. Science rarely, if ever, conclusively proves.

Similarly, one U.K. biology lecturer[22] claimed that science is an ever-changing process influenced by the curiosity of researchers. She described it as "a constant evolution" and claimed,

> The aim is to find out, in the most objective way—which is really a tricky word to use because as we see, there's not much objectivity to find out . . . [about] functioning or about knowledge . . . about how we work, how we are. And that's . . . I think, driven by curiosity.

For this scientist, the core of science is continued inquiry. She doesn't define science by its objectivity or ability to find "facts" or "truth," because, as she elaborated,

> This kind of evidence changes . . . so some things stay while some others just . . . they don't stand the test of time, but that's just how science works. It works on constant reevaluation of knowledge.

These scientists view science as a dynamic field of study and knowledge. It is not static, but changes with each new discovery. This steady change is driven by curiosity and the desire to know more. Within these descriptions of science is a sense of intellectual humility—an openness to new ideas and information, and a willingness to reconsider beliefs and recognize they might be wrong. As Merton writes, "Humility is expected also in the form of the scientist's insisting upon [their] personal limitations and the limitations of scientific knowledge altogether"—adding among his illustrations that the famous early astronomer Galileo taught his students to say, "I do not know."[23]

Some of our atheist scientists think that this view of science as dynamic and open to new knowledge necessarily implies an opposition to religion. Religion, they think, is steadfast in its beliefs and convictions, closed to questioning and new ideas, and resistant to change, challenge, and opposing views. But sociologists of science have questioned whether scientists are as open-minded as they publicly profess. Ian Mitroff's study of Apollo moon scientists, for example, suggests that they observed a norm of organized dogmatism, in which a researcher should believe in his or her own findings "with utter conviction while doubting those of others with all his [sic] worth."[24]

Science Defined as "Different" than Religion

Some atheist scientists, in their explanations of what science is, explicitly differentiated science from religious belief. One biology graduate student in the U.S.[25] told us:

So you can be 95 percent certain about it [a scientific finding] and that's pretty damn good, but you still recognize that there is a level of uncertainty that is possible and recognizing that uncertainty is part of science, whereas religion you're just 100 percent certain about these things because God presented them to you, but science is questioning that and saying, well, there is this off-chance that it's not the case and I can do an experiment and test that and show that it's not that by doing this.

Another U.S. graduate student in biology[26] defined science as the use of the scientific method to test hypotheses and then similarly emphasized that, "So whereas religion you're often told specific things, you're always questioning things with science." For these biologists, science is inherently different than religion in its presuppositions. As they see it, scientists can test theories and disprove ideas, and thus learn to embrace uncertainty. Religious believers, on the other hand, accept their beliefs on faith and do not allow themselves any room for doubt or testing. The doubt that drives hypotheses, productivity, and experimentation in science is seen as taboo for religious people. When we heard scientists distinguish science and religion, we often encountered their understanding of what religious belief is and what it involves.

Another way that atheist scientists contrasted science and religion was by framing science as something you do, as opposed to something that you believe. Doing science involves a constant practice of questioning, followed by research that finds empirical evidence to answer those questions. In contrast, these atheist scientists argue, religion involves believing in and thinking about the same unchanging truths for millennia. These scientists then would find it silly to say that you "believe in" science, and, on the flip side, to say that you "do" religion. (As an aside, Elaine finds this ironic, since there is an entire field of scholarly studies of religion that focuses on "lived religion," which examines all the ways people talk about "doing" or "showing" faith.)[27] This explicit emphasis on active process versus static beliefs can tell us a lot about the ways that some scientists view both science and religion. For example, one graduate student in physics[28] emphasized that science is an active process, saying:

Science—it's the study of how the universe works really. That's what I go by. So I don't think it's a belief system; it's just *a* [respondent emphasis] practice. It's a method for how to analytically look at what's going on around and try and understand it.

A professor of biology[29] told us when defining science:

If I didn't know about science, I could see that religion could provide that context of why things were happening. . . . If there was no science, and I was living 2,000 years ago, I would be religious.

For this scientist, religion does not exist in collaboration with or alongside science as a legitimate form of knowledge, but rather as an alternative worldview for the less educated and less reasonable.

Such narratives provide a small window into why some atheist scientists are anti-religious while others are not. There are atheist scientists who see a vast accumulation of scientific knowledge and question why scientific facts have not displaced faith altogether, while other atheist scientists recognize that most of humanity shares a desire to make sense of the world and some people find that understanding in science, some in religion, and some in both. Scientists can, and do, comprehend the quest for meaning and understanding that undergirds spirituality and religion.

Rather than demarcating science from religion, spiritual but not religious atheist scientists tend to define science in ways that implicitly overlap with religion. In such narratives, atheist scientists describe science as an endeavor that enhances the human experience. For example, when asked to define science, one U.K. biologist[30] stated:

Science is the aspect of humanity that is there to serve humanity in both gain of knowledge and, through that, increased understanding of the mechanisms by which the universe functions.

In this scientist's mind, the ultimate way to enhance the human experience is to increase human understanding of how the universe works; to do so will make human lives better. Previous research on scientists and spirituality,[31] some of which we discussed in the last chapter, has shown that scientists who express or identify with some form of spirituality tend to view science as inherently spiritual or feel that science can serve individuals in a spiritual way. A physicist from the U.K.,[32] when asked about his definition of science, explained his view in a way that sounds similar to how religious believers view religion: as a way for people to provide meaning, understanding, and stability to their lives. He said:

> I mean it's [science is] just the quest for an understanding of the world around you. Because a lot of people do have that need to understand what it's all about and I think it's trying to untangle the sort of incredibly complex and amazing world you see around you, but in a way that is predicted—yeah, that enables you to predict. I guess prediction isn't quite—it's understanding of it, but it leads to the ability to predict things and the ability to have more control over your own life.

Likewise, a physicist in the U.K.[33] defined science as inherently intertwined with the interests of humanity. She told us:

> Science? [*laughs*] Hmm, I suppose one can, you know, to me it's probably a very academic analytical study of the natural world and phenomena related to the natural world or how humans manipulate, if you want to say, the natural world to their own, you know, to their own benefit, usually.

From some spiritual but not religious atheists, we hear that science is both intrinsically and extrinsically valuable. Science is intrinsically valuable in the pursuit of fundamental truths about the world, even without application. The extrinsic value of science emerges when discovery leads to technologies and applications that enhance humanity, help people, and solve contemporary problems.

The Limits of Science

Are There Limits to What Science Can Explain?

We have seen how—through their definitions of science—atheist scientists rhetorically draw boundaries between religion and science. While they acknowledge overlap—mainly in the central quest to understand the world—they distinguish science from religion primarily by asserting the power of the scientific method over faith alone. The power that atheist scientists attribute to the scientific method raises the question: Do they see limits to what science can explain?

New Atheist scientists seem to have very definite ideas about the limits of science. For example, Richard Dawkins and Sam Harris often make provocative statements as they engage in their own kind of boundary-making. For instance, Dawkins posits in his book *The Virus of Faith*:[34]

> So atheism is life-affirming, in a way religion can never be. Look around you. Nature demands our attention, begs us to explore, to question. Religion can provide only facile, ultimately unsatisfying answers. Science, in constantly seeking real explanations, reveals the true majesty of our world in all its complexity. People sometimes say "There must be more than just this world, than just this life." But how much more do you want?

Scientists who hold views like this think that in all spheres of life, knowledge is only reliable if it is found through science. Likewise, for them, only questions answerable through science are worth exploring. Questions concerning the meaning of life are not even worth asking because they do not meet the criteria of questions answerable by science.

A small number of the U.S. scientists Elaine interviewed as part of her past work[35] were part of a tradition of scientists who think that science will make faith irrelevant. These scientists think that it is only a matter of time before science completely replaces religion. They claimed that religion has simply tried to answer the wrong questions in the wrong ways. As science continues to make further advances in the pursuit of knowledge, they reasoned, it's going to be harder and harder for religion to have a place in society. Those who adhere to this position of unwavering conflict between science and religion hold religion under the lamp of what they see as empirical reality and, in this light, faith is little more than a dim shadow cast outside the light of the lamp.

But our goal here is to dig deeper, to go beyond the New Atheists and ask how other atheist scientists see science and its connection to religion.

Atheists are often accused of embracing something called "scientism," a term which refers to the conviction that the scientific method is the only reliable way to understand the world around us. Often used in pejorative terms, scientism also represents an accusation that scientists have an exaggerated confidence that there are no limits to what science can explain. Sociologically, scientism is a boundary dispute between atheist scientists and subsets of philosophers, theologians, and other humanities scholars who assert that principles such as morality, love, or belief in God are incredibly important, but not the domain of the scientific method. Scientism emerged in part from 20th-century advances that transformed how scientists understand the world and emboldened their confidence in what the scientific community could achieve. In the wake of such progress, science popularizers—and New Atheist scientists in particular—have made various types of statements that have encouraged the "scientism wars" in the public sphere.[36]

We know that religious individuals in and out of science embrace both science and religion for understanding and experiencing the world. Even groups who are caricatured as "anti-science" accept a majority of scientific claims, though some are skeptical of scientists. But given that atheists epistemologically reject religion, we should expect a broad embrace of scientism among atheist scientists.

Of the 81 atheist scientists we spoke with for this book, we asked 50 about their views on the limits of what science can explain. Of these, only 17 responded in a way reflecting scientism. Modernist atheists (13) were the primary proponents of this view, with three culturally religious atheists and one spiritual atheist concurring. Thus, among the atheists that we spoke with, a majority actually *reject* scientism. And while modernists were the most likely to embrace it, they were almost split down the middle in their support or opposition to the idea (13 of 24 agreeing). Here we turn to what those atheist scientists had to say about the limits of science (or lack thereof) and the rationales behind such views. We begin with different dimensions of scientism we heard before turning to the views of scientists who reject it. Each type can be seen as attacking the legitimacy of religious faith as a means of understanding the world.

General Scientism

The most general form of scientism asserts that nothing is beyond science. In 1966, British biophysicist Francis Crick argued "the knowledge we have already makes it highly unlikely that there is anything that cannot be explained by physics and chemistry." Outspoken British chemist Peter Atkins stated in the early 1980s that "there is nothing that cannot be understood, there is nothing that cannot be explained, and everything is extraordinarily simple."[37] In the same era, world-famous theoretical physicist Stephen Hawking described the goal of physics as "nothing less than a complete description of the universe we live in."[38] One strain of this general scientism asserts that the *only* legitimate domain of truth is that which stems from natural scientific knowledge. In recent years, atheist science popularizers have advanced the notion of scientism in the public sphere while philosophers have debated the forms and validity of the idea. Popularizers such as Hawking, Neil deGrasse Tyson, and Bill Nye have made similar statements attacking philosophy, unsurprisingly drawing accusations of scientism from the philosophy community.[39]

Of the atheist scientists we spoke with who embrace scientism, many were relatively general but nevertheless confident in their view that science can explain everything. They were quite assured in answering our questions about the limits of science. Scientists who offered only general statements on scientism did not mention religion, ethics, or morality, but their narratives imply two things. One is that religion could be explained by science, given that science can explain everything. The other is that science, and natural science in particular, is the only legitimate way to understand the world. One biologist in the U.K., a modernist atheist,[40] quickly answered, "No" when asked, "Do you think there are limits to what science can explain?" She continued:

> Whether some things are remotely close to being able to be explained is another question. Whether the scientific method of exploring the universe can potentially address all questions? Unequivocal yes.

Another modernist biologist in the U.K.[41] gave a similar response when asked about the limits of science. In his words:

> I think there are limits to what science can explain right now because we don't have a full appreciation of what goes on around us. I think science will eventually give us at some point a full understanding of what's going on everywhere, OK? Whether it be a physical sense, a chemical sense, or biological.

These scientists illustrate that scientists can exercise humility in the current state of scientific knowledge while offering unabashed confidence in what is possible.

One modernist atheist scientist who works as a postdoctoral fellow in physics in the U.K.[42] took a slightly different tack when describing the lack of limits for science. When he was asked if there are any limits to what science can explain, he remarked that science may even be able to explain life after death under certain conditions:

> Well, maybe not. Maybe eventually yes, who knows? I cannot exclude the possibility. Yeah, probably not. Probably if—but, who knows. Anyway the thing is that I think it's the best way of investigating the world we have. So if science cannot give us an answer, nothing else can in my opinion.

Epistemic Scientism

A second form of scientism is epistemic scientism, which asserts that if science cannot grasp something, it is not a part of reality. There is an emphasis on materialism—the basic premise being that only atoms or material particles exist in the world. Epistemic scientism is captured by Crick, who said, "You, your joys and your sorrows, your memories and your ambitions, your sense of personal identity and free will, are in fact no more than the behavior of a vast assembly of nerve cells and their associated molecules."[43] It is reflected in Carl Sagan's statement: "I am a collection of water, calcium and organic molecules called Carl Sagan." (An experiment for the reader: At your next social gathering, introduce yourself in this manner and see what happens! We sociologists typically go with, "I am nothing more than a product of my environment.")[44] In other words, consciousness, identity, and free will are all the product of the "molecular machines"—in Sagan's terms—that comprise our bodies. In this view, science provides a complete account of human beings and the universe. If something is not explained by science, it does not exist. As philosopher Alexander Rosenberg puts it, "being scientistic just means treating science as our exclusive guide to reality."[45]

Comments from a modernist atheist physicist in the U.S.[46] illustrate this perspective on the limits of science. She said:

> I don't think there's anything beyond science. I do think there are things that I'm not sure . . . just because they're so complex systems. I'm not sure science will ever be able to tell me what you're thinking right now, not because it's beyond science or there's some kind of mystical thing, but more that it's such a complicated system that it can't be reduced to the, you know, mathematical sum of its parts.

The epistemic emphasis in this narrative can be seen in the dismissal of a "mystical thing" and the idea that only a math-oriented method can confirm what is real or possible. In short, something needs to be within the scope of science for it to be considered reality.

When we asked a modernist atheist biologist in the U.K.[47] about the limits of science, she responded:

> I think that in our physical world, the physical, chemical, biological world, there aren't limits. Eventually we'll understand most things. The most challenging thing is our consciousness. . . . That is associated with spirituality and religion, and I don't feel those are things that science needs to explain, because they're made up by us. They are the product of human consciousness, human needs, and they serve human needs, but they're not entities in themselves, they're products from us.

This narrative very clearly illustrates epistemic scientism in the assertion that there is no reality outside of the natural, material world, and all things associated with "the sacred" are not real entities and thus do not merit explanation by science. On the one hand, science has no authority or need to explain things outside the natural world. At the same time, it is believed that the only valid knowledge comes from science. The only reality that we can know is the reality that science can reveal.

According to epistemic scientism, our joys, sorrows, ambitions, and other feelings are outcomes of our molecular makeup. Crick famously referred to this as "the Astonishing Hypothesis"—noting that the view is "so alien to the ideas of most people alive today that it can truly be called astonishing."[48] We should be careful, however, in drawing the erroneous conclusion that epistemic scientism leads scientists to a meaningless or uninspired view of the world. Even "bag of molecules" Carl Sagan stated, "I find it elevating that our universe permits the evolution of molecular machines as intricate and subtle as we are."[49] This ability to find inspiration and meaning through science exists among the epistemically scientistic atheists that we interviewed. For example, in our discussion about the limits of science with a modernist atheist biologist in the U.K., she explained that, for her, understanding how a phenomenon might be reduced to its constituent parts does not reduce the phenomenon:[50]

> If [love is] just chemicals in our bloodstream, which is probably what it is, then it's kind of neat to have that understanding. . . . I like understanding the fundamentals of something and being able to explain—so it doesn't take away the beauty of love or the meaning of love to me. It's just getting a deeper understanding of it.

The Limits of Technology (At Least for Now)

A majority of the atheist scientists we spoke with do see limits to what science can explain. The most prominent narrative among these atheists is that science is bounded by the tools available. Very few would say that we currently possess all the necessary tools and theory to give an account for everything in the universe. For example, when we asked a lecturer in biology in the U.K.[51] if she thinks there are limits to science, she responded:

> Yes, I do. But I don't think that means we need to invoke some kind of religious explanation. I just think it means we have to accept that there are some things that are really big and really hard and we don't have the tools to answer those questions.

In many respects, attributing the limits of science to technology can be viewed as a "soft" form of scientism. Such a position effectively argues that science *could* explain everything *but* cannot at the present moment because our technologies are limited. This is a different position than arguing that certain questions, no matter what technology is available, cannot be answered by science.

Indeed, many atheists who believe technology limits the explanatory power of science see the situation as only temporary. They believe that—given enough time—technology will one day advance to the point where everything is explainable by science. Many scientists, especially those who are modernist atheists, have a sense of inevitability—that we will eventually possess all the tools we need to discover anything and everything. This sense of inevitability came up in our interview with a professor of biology in the U.K.[52] who told us frankly:

> Throughout history, there have been many times things have been unexplained, and science wasn't able to explain it. . . . I guess examples like thunder used to seem as the god Thor being angry or something like that and people think that it's magic—but as science progresses and tools are developed to investigate, some of the things that are mysterious become clarified and one understands the fundamental basis of these, and that is, I guess, what science is about, and that's why it changes.

Another U.K. atheist scientist[53] said something very similar when asked if science has limits:

> Uh, personally probably not, but . . . you are limited by technique rather than what you can explain. So you have limitations, but the limitations are, you know, they are dependent on what you can observe and what you can measure. Which means that where you are now, you may certainly have limitations in how you interpret what happens around you. But that's not to say 10, 20, 50, 100 years from now, that the same people, effectively, doing the same types of experiments but with improved methods, or through detection or observation, will alter. So I will say theoretically no, I don't think you're limited. But in practice you're limited by where you are in terms of the development of the methods and what you have at hand.

This, again, is the sense of inevitability. Our abilities are limited by our technologies now, but that is not to say that we will always be limited in those abilities. Moreover, given the progress of science throughout history, it is likely that we will one day overcome the current limitations and obstacles to scientific discovery.

For instance, one modernist atheist biologist from the U.K.[54] believes that, as long as humanity can stave off catastrophic destruction, what we can learn through science for the betterment of humanity is essentially limitless. She said:

> There are certainly limits in every day and age. I think those limits keep constantly being pushed forward so what's limited today may become possible in the future. I don't know if mankind will destroy itself before it ever—I don't think there's ever a limit anyway, so I think there will always be more and more to understand. So in a way I would say it's limitless what there is to

understand, and science is always pushing forward. But in each time it's limited, yes.

The same modernist atheist scientist from the U.K.[55] also expressed the view that there is no limit to the future of science. She told us:

> Looking at the history of science I've seen that often the best scientists are the ones that question dogma and I've seen, you know, theory change or be refined or added to or become more global. You know, history tells us that . . . what was a limit before ceases to be a limit at some point.

Human Limits Limit Science (At Least for Now)

While some scientists think that science has only technological limitations, other scientists take a slightly different approach. Science, they believe, is limited by human cognitive powers. Again, science *could* provide all the answers, but will reach a limit because there are features of the universe we will never be able to understand given the abilities of the human mind. In other words, some questions may go unanswered, not because of limitations in science, but because of limitations in being human.

Of the atheist scientists we interviewed, 12 mentioned such limits; this included two modernist atheists, seven spiritual atheists, and three culturally religious atheists. One physicist from the U.K.[56] put it like this:

> I think we're only scratching at the surface because as humans we've got a very limited ability to think in multiple dimensions. I mean, the whole concept of time; what is time? . . . Is it just something that comes—is it intrinsic or does it come about—or—or

[at a loss for words]. It's a complete enigma to me, that, yeah, I think we're scratching at the surface so there's an awful lot out there that we find difficult to understand because of the limitations and the way we think and the fact that we're sort of rooted in three dimensions in time and physical bodies.

A biology professor from the U.K.,[57] a culturally religious atheist, said similarly:

Yes, but I don't think the limitation is the scientific method. I think the limitation is the fact that science is only done by humans. But I think the method itself has the power to explain anything given a large enough intelligence. We don't have the intellectual capacity to generate models of sufficient complexity and subtleness to explain much of what we've seen. I mean we do have some advantages, but . . . I think there are things that are beyond our limits. So, for example, complex behaviors of ocean currents and things like that—we can make first and second order predictions about them, but there's just too much detail, too much noise, too much detail.

Likewise, a graduate student from the U.S.,[58] who is also culturally religious, contrasted our ability as humans to the abilities we would need to understand everything in the universe:

Assuming we can invent anything we want, yes. Assuming we can, like, bend the laws of physics to do anything we want in the universe and go anywhere we want and measure anything we want, yes. Theoretically, I mean. Using methodology, [can we] investigate truths? Sure. If we can do anything, then we can. If we are gods ourselves and can just float around and do anything we want and look at things and report things down then, yes, we can answer every question in the universe. Will we be able to do that? I do not think so.

For this scientist, we would have to be "gods ourselves" to discover and comprehend everything that there is, to perceive all of reality, and to apply science to its utmost potential.

Meaning-Making, Morals, and Other Limits to Science

There is another type of scientism that entails the conviction that science can answer questions related to meaning, values, and ethics. Perhaps inspired by new advances in neurobiology, biologist Edward O. Wilson stated in 1975 that "science may soon be in a position to investigate the very origin and meaning of human values from which all ethical pronouncements and political practice flow."[59] More recently, in his book *The Moral Landscape*, Sam Harris argues that "science can, in principle, help us understand what we *should* do and *should* want."[60] On the one hand, Harris makes nuanced arguments and makes clear that he is not contending that science can provide a biological account of what people do in the name of morality. On the other hand, the subtitle of his book is "How Science Can Determine Human Values."

For some atheist scientists, however, questions related to purpose, meaning, and values fall beyond the reach of science. This view of the limits of science fits nicely within Stephen Jay Gould's theory of "non-overlapping magisteria," which posits that questions of science (based on facts) and questions of religion (based on purpose, meaning, and values), occupy different domains, and are therefore both legitimate and not in conflict.

Eleven atheist scientists we interviewed—four modernist atheists, two spiritual atheists, and five culturally religious atheists—discussed these kinds of limits on science. For example, one modernist atheist graduate student[61] from the U.S. told us that he finds certain questions ill-suited to having scientific answers:

Is every question conceivable ever going to be answered? . . . Such as, does God exist? Or what is the meaning of life? In my opinion, they are questions that are ill-posed in the sense that using logic or kind of a scientific method to answer them definitively yes or no is just not practical.

Another U.S. physicist,[62] who is culturally religious, agreed that not all important questions are answerable through science. She also pointed to questions of meaning and purpose, saying:

I think there are certain kinds of questions that are ill-posed in terms of having scientific answers. Like, why is there a universe? That's not a question that science can answer. But I think it's just ill-posed. Why does there need to be an answer to that question? I don't know. Having to rely on an answer to that question to give yourself meaning is easy but not ultimately useful, I think.

This scientist argues that there are not only questions that science cannot answer, but also questions that do not need to be answered by any means. Instead of turning to religion as a way to fill in the gaps left by science, this scientist is comfortable with the unknowable and the ambiguous.

Other scientists expressed the idea that we will never be able to use a scientific approach to answer the deepest mysteries of the world or solve all problems. For example, a culturally religious graduate student in biology[63] argued for limits to science and a more humble approach to the idea that science can provide an answer to everything:

I don't know, but I do—I don't think science can explain everything. . . . There are always going to be questions we can't answer, but am I uncomfortable with that? No. . . . Why would we think that we can answer every single question in the universe? It's kind of a bit of hubris in my opinion. . . . OK, here's a question: How do we make world peace? Everyone knows the answer. Can we do it? No. Do we need science to tell us? No. The answer is very straight-forward. We still can't do it. So there are many questions that are answered that we still can't do and we'll never be able to do in life.

Next we turn to a discussion of how atheist scientists find purpose, meaning, and morality in the midst of their atheism.

7

How Atheist Scientists Approach Meaning and Morality

"Do you know what a virus looks like?" a British physicist[1] asked us in response to the question, "How do you deal with questions about the meaning of life, such as 'Why are we here? What's the purpose of my life?'" At first it seemed to us like he was evading the question. He continued: "You can specify a virus by the positions of the atoms in its structure. And that's been done for many decades. Lots of viruses are specified simply at the molecular level. And in some sense, we think of them as being alive. In what sense are they alive? They're alive because they have a structure, which permits them to—[in] a particular set of circumstances—self-replicate. And as far as I can make out, that's the only distinguishing feature of life."

Intrigued, we pressed further. How did this view of life, we asked, provide him with a sense of purpose? "I don't think there is any sense of purpose" he replied. "I don't see [that] there's any necessity for a sense of purpose."

How do atheist scientists grapple with questions of meaning and purpose? Many people might expect most atheist scientists to share the view that there is no ultimate purpose in one's life, and that this (lack of) belief makes for a dreary existence. In contrast, philosophers who trace the history of ideas tell us that, historically, the Christian religion and the other monotheisms have offered the simplest explanation to the question of purpose: life has inherent meaning and purpose because humans are made by God and in the Image of God (what some Christians call the Imago Dei).[2] Religion's monopoly on explaining the meaning and purpose of life weakened

Varieties of Atheism in Science. Elaine Howard Ecklund and David R. Johnson, Oxford University Press.
© Oxford University Press 2021. DOI: 10.1093/oso/9780197539163.003.0007

somewhat with the Protestant Reformation, the Renaissance, and the Scientific Revolution.[3] These movements did not necessarily foster atheism, but changed the basis for thinking about the meaning of life and led some individuals to question the veracity of religious perspectives on meaning and purpose. With the rise of Darwin's theory of evolution in the late 19th century, secular perspectives on the meaning of life—in particular, the nihilist view that there is no meaning *and* that questions about meaning may not be worth asking—began to infringe on religion's monopoly.[4] In her article "Atheism and the Meaningfulness of Life" in *The Oxford Handbook of Atheism*, the philosopher Kimberly Blessing writes:[5]

> Since so many people are concerned with living meaningful lives, it would be a great coup for one or the other side if it were able to lay exclusive claim to meaningfulness. Hence in their attempts to win over converts both theists and atheists attempt to show that their opponent's orientation towards religion prevents them from living truly meaningful lives. But exclusivists on both sides are wrong. For neither atheists nor theists are *necessarily* committed to meaninglessness. . . . [And] neither atheism nor theism precludes meaningfulness.

Similarly, sociologist Jacqui Frost has studied new ways that atheists articulate meaning and purpose and found that some atheists have begun to reject the notion that they need certainty in their unbelief in order to have meaning. Instead, many move back and forth between certainty and uncertainty, all while finding purpose and meaning *in the uncertainty itself.*[6] At the same time, Frost has studied atheists who participate in church-like organizations like The Sunday Assembly for the express purpose of meaning-making and social construction.[7]

Other sociologists have examined the relationship among meaning, religion, and secularity. For their part, the central focus has been on "meaning systems," constellations of beliefs about the world, what people believe exists and does not, and how groups—particularly religious groups—uphold particular versions of meaning systems.[8] Some sociologists compare the ways that religious and nonreligious individuals construct meaning. For his book *Living the Secular Life*, for example, sociologist Phil Zuckerman interviewed secular individuals about their attitudes toward a wide range of moral issues and thoughts about the meaning of life.[9] Much more frequently, however, sociologists use survey questions about belief in God and the relationship between science and religion to identify the role religious belief, atheism, and science may have in helping individuals derive meaning from their lives.[10]

Here we ask how atheist scientists respond to questions about meaning and purpose. Do they even think such questions matter? Are they a bunch of unhappy nihilists?[11] How do they create meaning in their day-to-day lives? Where do they find their purpose? What might their approach to meaning and purpose tell us about their approach to morality? These questions are important to explore because many people distrust atheists and believe they are morally depraved[12]—perhaps because we as people believe we behave better when we think there is a God watching us, according to research by psychologist Will Gervais and his colleagues.[13] Understanding how atheist scientists answer these questions may address public stereotypes about atheist scientists.

However, we actually found that the public is somewhat correct in its perception of atheist scientists as nihilistic, partially confirming these stereotypes. When we asked questions of meaning and purpose, we found that, among all atheists, there was an overwhelming emphasis on the idea that life has no meaning. In other words, a nihilistic position was the primary tendency among modernist, spiritual, and culturally religious atheists. However, it was not the only position. Another common viewpoint among atheist scientists was the belief that questions of meaning cannot be answered at all, an important distinction that indicated that meaning *could* exist, but that it simply cannot be found. Still another group posited that meaning can be found in the progressive nature of science, in discovering more for the greater good. Finally, others, particularly those spiritual atheist scientists we met in Chapter 5, *did* tell us that transcendental meaning and purpose can exist in tandem with their atheism. These positions continue to underscore the nuance that exists among atheist scientists.

"Biological Accidents with Consciousness"

As we stressed in the previous chapters, atheist scientists are not a homogeneous group. We did find some unique tendencies among the different types of atheists in how questions of meaning and purpose are addressed, but these differences were a matter of degree rather than deep substantive divides. For example, nihilism, which generally refers to life without meaning, "was the creed of most of those who rejected religion throughout the nineteenth and twentieth centuries," British philosopher John Gray notes in his book *Seven Types of Atheism*.[14] "[It] remains the creed of most secular thinkers today," he writes.

We found a nihilist view was popular among the atheists in science we interviewed.

Most atheist scientists we spoke with believe there is no ultimate meaning of life. When asked whether he finds questions about meaning and purpose important, one spiritual atheist biologist[15] told us:

> I currently firmly subscribe to the idea that there is no meaning of life. There is no purpose. And I think it's an idea that's been evolving in my head all my life—that we are biological accidents with consciousness. . . . And that puts us in a terrible predicament; because we are self-aware and we are aware of the universe around us. And because we are self-aware, we want to find a purpose for ourselves. But I don't think there is one. . . . And this is the huge paradox that we have to come to terms with; that we just happen to be here for a limited time, and then we have to get

over it and then we're out. The only thing we really leave behind is some memories and some children. And some manure.

Not all atheist scientists find questions about meaning and purpose important. Many scientists we talked with implicitly or explicitly indicated that they find such questions uninteresting or irrelevant. One culturally religious atheist[16] explained why he finds these kinds of questions trivial, saying:

> They're not important to me at all. And it's weird to say that, but I really just don't care. I don't. I really just don't care where we came from. We came from something . . . .I'm perfectly content that it's a mystery. I don't need to know the answer to everything, and it's funny because religious fanatics need to know an answer, right? There's an answer for this and it's in the Bible, right? Well, scientists who are fanatics are like, we will eventually know the answers to every question through science. . . . I'm happy that there are still things we can't answer; otherwise, the world would be a really boring place.

This scientist's response is interesting. On the one hand, he asserts that questions of meaning are unimportant, yet, on the other hand, he assigns value to mystery and curiosity, which often flow from questions of meaning. It is also interesting that, in expressing his opinion, this scientist critiques both biblical literalism and scientism, along with the idea that ultimate answers to questions of meaning can be found in either religion or science.

A modernist atheist scientist[17] told us, "I'm never really sure why people have to try and find a meaning of their life." In her view:

> There've been hundreds of thousands of species before us and there will be after us and we are a relatively small and probably not very successful blip in . . . the context of the earth or the universe. And so I don't really feel like I'm seeking some kind of specific meaning in my life.

Many scientists referenced the temporality of human life in explaining why they view questions of ultimate meaning as unimportant, including another modernist atheist scientist[18] who said, "We have been on this planet for a flicker of the whole existence of time. It's completely insignificant. Humans are naturally solipsistic, OK?" When we asked one atheist scientist "why are we here?" in the cosmic sense, she answered, "It's random chance."[19]

Atheists we spoke with were particularly critical of those who attempt to place humanity at the center of everything in order to generate a veneer of meaning and purpose. Instead, they explain our existence within the scientific framework, referencing natural processes like the Big Bang and evolution. In their view, humans occupy a miniscule place in the vast history of our universe and billions of years of genetic progress. We have no greater meaning or purpose than any other species. For example, one culturally religious atheist[20] explained:

> I view it as a very mechanistic sort of issue. . . . We are the product of an evolutionary process, which started an awful long time ago, many of the beginning parts of which we don't understand very well once it got started. . . . I don't really think about it, but I don't think God was involved. . . . Even if you acknowledge there was an evolutionary process.—The meaning of my life is to carry my genes on into the next generation.

"Why am I here really? Because I'm part of a large biological infrastructure," a modernist atheist scientist[21] told us, similarly drawing on an evolutionary perspective. "And that it happened to arise purely by chance."

Existential Crises among Nihilistic Scientists

In the United States, 88 percent of both actively religious individuals and atheists report that they are either "very satisfied" or "somewhat satisfied" with life. When we break that down further, we find that 63 percent of actively religious individuals report they are "very satisfied" compared with 59 percent of atheists.[22] In England, a nationally representative survey of 7,000 individuals found no differences in a broad variety of mental health outcomes across religious and nonreligious individuals.[23] If we look more broadly at measures of societal well-being (such as life expectancy, literacy rates, child welfare, and wealth), nations that are relatively nonreligious, such as Sweden, Norway, France, and Britain, tend to score higher on most of these measures than more religious countries (suicide rates, however, are lower in religious countries).[24] In short, it is difficult to conclude that atheists are more likely than religious individuals to be unhappy or experience lower levels of well-being. Life without God can be meaningful, purposeful, and happy.

Nevertheless, big questions of meaning and purpose do seem to stir up an emotional undercurrent of pessimism and in some cases depression among the most nihilistic group of atheist scientists with whom we spoke. There are three patterns that we identified in such accounts. First, and most consistently across these accounts, some atheists acknowledge that a purely materialist answer to questions of meaning and purpose is not emotionally satisfying. As one modernist[25] atheist explained:

> I'm not sure there's any systems in life, unfortunately. I think it's just random chance. . . . I think there's no sort of grand purpose or anything like that. I don't think it's a very comfortable sort of vision, but it's just more of a realistic vision.

This scientist concedes that believing that life does not have an overarching purpose, the only option he views as realistic, is less comfortable than believing in a predetermined existence. We also spoke with atheist scientists who find no purpose in the rituals that provide comfort to others. Their pessimism, they say, stems from an inability to see any function to rituals to which others assign value. We spoke with one modernist atheist,[26] for example, who, when asked how he answers questions of meaning, responded:

> You try not to think about that. It's actually depressing. . . . Last week I was in India, actually. I was in a city [Varanasi] on the Ganges [River] where the Hindus burn their bodies. . . . The Hindus believe in reincarnation, but if you die in Varanasi and then you burn . . . you go straight to heaven, so you skip the cycle. They had all these people making these funeral piles on the Ganges. And the whole thing was so alien to me, but hundreds of thousands of people who believe strictly, they go to Varanasi to die. For me, they're utterly wrong. It doesn't mean anything. . . . It really felt alien.

A second dimension of the "pessimistic" perspective on meaning and purpose is compartmentalization. In the mid-eighteenth century, Scottish philosopher and skeptic David Hume described how human psychology, distraction, and compartmentalization protected him from despairing over life's big questions. "Where am I, or what? From what causes do I derive my existence, and to what condition shall I return?...I am confounded with all these questions and begin to fancy myself in the most deplorable condition imaginable, environed with the deepest darkness, and utterly deprived of the use of every member and faculty," he wrote in *A Treatise of Human Nature*. "Most fortunately it happens, that since Reason is incapable of dispelling these clouds, Nature herself suffices to that purpose, and cures me of this philosophical melancholy and delirium, either by relaxing this bent of mind, or by some avocation, and lively impression of my senses, which obliterate all these chimeras. I dine, I play a game of backgammon, I converse, and am merry with my friends. And when, after three or four hours' amusement, I would return to these speculations, they appear so cold, and strained, and ridiculous, that I cannot find in my heart to enter into them any farther."[27]

Our interviews revealed that some atheist scientists find their answers to ultimate questions emotionally unsatisfying and, as a result, seek to avoid them. Atheist scientists we spoke with described demarcating their day-to-day life from matters of meaning and purpose as a way to protect themselves from struggling or stressing over these questions, though compartmentalization can sometimes be hard. "I don't think about it a great deal," said one modernist[28] atheist scientist, when asked about ultimate meaning and purpose. "I think I would be too terrified if I really stopped to think about it."

Some atheists we spoke with also indicated that they had experienced depression for a sustained period in large part due to lack of answers to questions of meaning and purpose. When we spoke with one modernist atheist scientist,[29] she told us:

> I've actually struggled with depression for quite a number of years and sort of, "why am I alive," has come up a lot. I've never found an answer. I just don't know, but I'm here and I have to believe that there is a reason to wake up every morning. So not knowing what that is, I get up every morning and I come to work. Even if there is not a reason then I would still like to live comfortably because I am here.

Other atheist scientists described past episodes of depression. For some, these coincided with abandoning religious faith. One modernist atheist scientist,[30] for example, told us that:

> The big questions are unknowable. And thinking back to those sorts of problems I had, as I lost my faith, was this shrinking down of the universe to human dimensions.

A culturally religious atheist[31] said about questions of meaning and purpose:

This is an area in which—as a scientist—it can easily become slightly depressing because the real answer, as far as we can tell, is that there is no meaning. And there is no purpose. And some scientists have written quite eloquently on that subject and it can, it can make quite depressing reading.

"Do *you* find it depressing?" we asked. This physicist went on:

No, not now. . . . As a younger man . . . [I] found [in] those kinds of issues some sort of a kind of emptiness, which was uncomfortable.

Yet, even though some atheist scientists may exhibit patterns of nihilism and pessimism when asked about big questions of meaning and purpose, we see that others take different approaches to these questions and philosophies. Some atheists, for example, don't think that questions of meaning and purpose will ever yield an answer, and it is in the unknown that they take comfort.

Meaning Is Not Answerable

Other atheist scientists believe that we cannot answer questions related to meaning. This is not to say meaning does *not* exist, but rather that it is beyond our comprehension. When we asked one modernist atheist scientist[32] about big questions related to meaning, she replied:

I don't have any answers to those questions. That would be nice to answer those questions. . . . [But] that's a question science won't answer . . . . I kind of feel that there may be no meaning. It may be just that the universe is a thing, it's a pattern. It could be one of many universes for all we know.

For many atheist scientists, questions of ultimate meaning and purpose belong to the realm of religion, which they see as totally separate from the realm of science (evoking Stephen Jay Gould's notion of non-overlapping magisteria, in which science and religion are totally separate, addressing different types of questions). In the U.S., 38 percent of atheist scientists believe that science and religion are independent of one another and refer to different aspects of reality. In the U.K., 36 percent of atheist scientists embrace this view. "I don't know that I can approach it," said one spiritual atheist[33] about big questions of meaning and purpose. "I think that's probably what religion has evolved partly to do, but I don't feel that I need to answer that question and I don't know that I can. I don't know that humans have ever been or had the ability to answer that question." A modernist atheist scientist[34] made a similar point and added: "I don't think just because you can't answer 'what is the meaning of life?' that I'm a nihilist and I don't think that there's any meaning to life."

More than Nothing

We are also able to identify distinct ways in which some spiritual atheist and culturally religious atheist scientists discuss meaning and purpose in life. They are more likely than modernist atheists to draw from bodies of knowledge beyond physical and natural sciences to think about questions of meaning. We find that these scientists still reject sacred meaning and still assert materiality as above other things, but that they also embrace other faculties alongside rationality. "I suppose my reading of that comes from sort of existentialism, Buddhism, Taoism, things like this. There is no predestined—predetermined—purpose, but ways of painting a picture," one spiritual atheist[35] explained. He continued, relating "painting a picture," creativity, and purpose to each other. "Camus talks about creativity as being a good place to put your purpose. I think that's—yeah, that's an aspect of it [for me]." Another spiritual atheist[36] told us:

> I don't think there's a purpose, a particular purpose in my life. I just happen to be here and I have to go with it. I know that I'm a scientist because I'm curious about the world and understanding how it works. I think most scientists have kind of a sense of wonder of the universe. . . . In *The Tao of Physics*, [Fritjof Capra says] that when you meditate, you have that sense of belonging to the universe, you get a certain feeling . . . you get the realization of something. . . . This is what we're looking for [in science] and I think I kind of agree.

But others did find larger purpose, particularly in the teaching aspect of their work as scientists. They viewed their role as educators, with the ability to share information and mentor students, as one that allowed them to impact others for the better. For example, a spiritual atheist professor[37] told us:

> I can impact young people's lives. We're really shaping their lives, their future. . . . Even though you have a big discovery, that's—compared with the nature, that's very insignificant. But if somehow you can influence other people's life, I think that is really satisfaction.

This scientist acknowledges that her scientific contributions, put into perspective, can feel like they don't matter. Her impact on students, however, can make her feel like her life has value. This perspective was echoed in another scientist,[38] who told us, "Part of my purpose is teaching and then teaching in a way that lets other people find their paths and lets other people find their utility. . . . I derive a lot of meaning from those things."

Some religiously involved atheists were very direct in describing how they integrate religious ideas into their views of meaning and purpose. In formulating his sense of purpose, one culturally religious scientist[39] emphasized that he draws on Jewish thought, prioritizing commitment to the well-being of humanity over self-interest. He discussed meaning and purpose in terms of what he would tell his children:

> I think that science is not very good at some of the questions you just raised such as the purpose of life. . . . Regardless of the purpose of life, you are part of mankind. . . . Our own religious belief is founded on the view that all humans are related to other humans. That is the essence of the Jewish philosophy—that there is a single God and that is in the sense that our ultimate guide for living is a panhuman one. It applies to all humans equally. And that you, my children, have to be cognizant of not only your own comfort and survival but the comfort and survival of the community in which you live. . . . And eventually you will come to understand that your community is humankind.

For this scientist, his atheism ensures that purpose holds no divine connotations for him, but rather simple helpful actions, making the world a better place, and community all provide meaning and guidance for one's life. Other atheists held this attitude even without a cultural connection to religion. One spiritual atheist physicist[40] told us:

> I do try my best when it comes to work but also in interactions with other people, to not pollute, to try to look after what bits of land I affect, so my garden, like when I pull the vegetables, I try to make sure that I put down new compost that's similar to put back what I've taken, that sort of attitude . . . giving back what you take and trying to strive to grow as a person and be a better person I guess.

Like her peers cited earlier, this scientist is not adrift in a nihilistic swamp, but has found ways to articulate purpose in the midst of her nonreligion through her connections to the world.

Not Providence, But Progress as a Meaning System

If the universe is random, purposeless, and godless, where do atheists find direction? In his book *A Meaning to Life,* philosopher Michael Ruse considers two metaphors that contrast religious and secular paradigms of meaning and purpose. One metaphor, representing the religious perspective, is a providential view of life that emphasizes a God who has created the world, is in control of it, and shapes all meaning and purpose. The other metaphor—inspired by evolutionary thinking and Darwinism in particular—is one of progress, in which the cultural and biological worlds are continuously improving. In the words of Ruse: "Both Providence and progress are aimed at a good ending, but whereas for Christianity it all comes down to God, for the progressionist it is all a matter of rolling our sleeves and doing it ourselves."[41] In short, it all comes down to humans. In discussing purpose, atheist scientists often clearly categorize themselves as scientists and articulate a form of purpose specific to scientists and squarely rooted in this particular professional identity. For them, science motivates a purpose based on continuous progress. Sociologist John Evans concurs, arguing in his work that meaning in science emerges from striving for constant progress.[42]

Among the atheists we spoke with, the majority of those who stressed a progressionist view did not employ the language of evolutionary thought, yet the view of progress among many biologists is rooted in evolutionary thinking. One modernist atheist[43] biologist, for example, told us:

So being a scientist and having a Darwinian view of not just life, but the universe itself—I've read numerous books [by] the likes of Richard Dawkins, Sam Harris, Daniel Dennett, Christopher Hitchens. When you have the search for knowledge—not the acquisition of knowledge—the process by searching it, to me, gives meaning to your life enough, OK? So the process by which you can take a scientific approach to understand the world around you gives meaning and then to pass that knowledge onto your offspring.

When we asked one of our modernist[44] atheist scientists how he personally thinks about big questions related to the meaning and purpose of life, he initially responded, "I think they're fairly ridiculous questions in general—they basically can't be answered." However, he continued, stressing advance of science itself as providing him meaning:

Of course, I'm sort of an evolutionary biologist myself and so we're here as organisms that have a certain life span—we reproduce and we consume things and grow. And then there's the human side is of course a little different because we try to produce something, build something, provide something for the next generation, add to the science, add to the community.

One culturally religious atheist biologist[45] gave us a detailed view of the progressionist perspective, stressing that science enhances the world and is a very different source of meaning than religion. He said:

The meaning of life is nothing more than what you see in front of you. And that we are evolved beings [with] all these impulses and desires that come out of our evolutionary past. But . . . as Darwin put it . . . there's an impetus towards having a better life for everyone and trying to achieve a good life and actually contributing to those efforts is the common good. . . . Religion hasn't provided that. What we've had over the last 200 years since the Industrial Revolution and Enlightenment is that we've lifted humans out of squalor . . . we live for twice as long as we did and children don't die in childbirth. . . . I'm wed to the idea of progress—both political and scientific progress. Science is providing us with better healthcare and actually lifting people forward.

This scientist was able to take a bird's-eye view of progress, and the way that technological innovations have improved the human condition, and apply this perspective to the way he thinks about and lives his life on an individual level. He continued:

So the meaning of my life is that I'm driven by two things. As a scientist, one is driven by curiosity and a desire to know more about the world and find out about things and be surprised by what you discover. And the other is that it would be nice to say that . . . you've made some contribution to . . . those efforts to improve the human lot.

For many atheist scientists, the metaphor of progress anchors a secular meaning system in which science leads to advancements that transform and improve the world. "I find purpose in hopefully making big important discoveries that other people will be able to find useful when they make their big important discoveries," said one modernist[46] atheist scientist. Some atheist scientists discussed scientific progress in relation to "making their mark" through science. "I find *meaning and purpose* [emphasis ours] in making a difference that my life existed versus that it didn't exist," one modernist[47] atheist said. "I'd like to leave a mark by the relationships I have and the love I leave behind and also the creation in terms of work that I leave behind." The spiritual atheist scientist who discussed finding purpose in teaching[48] also emphasized purpose in making a mark as a scientist, telling us:

Part of my purpose is to do great science and leave it behind and have it be useful to other people decades from now. . . . Part of my life is reaching beyond the traditional gender role of mother . . . to do something else.

Most scientists, however, discussed progress in terms of materializing new knowledge and technologies that deliver benefits to society by enhancing health and human well-being and helping us understand and address societal and environmental problems. Such benefits to humanity provide them a sense of importance and drive. Some scientists discussed cultural progress and relational progress— changing and shaping the future lives of others through their work. Often, such accounts included a critique of religion. One modernist[49] atheist scientist, for example, said:

I think the meaning of life—being a scientist—I stick with questions I can answer. The question as to what my life is, well, I mean I do think society has an obligation to live together on a planet. And some of the biggest crimes have been created in our time are going to be seen and your children will see it . . . as a complete mess we've made of the world. And praying to God won't help you. If I was God, I would say: Look, I gave you brains, use it, because you're not.

There are three noteworthy aspects of this scientist's response. First, like other atheist scientists, he firmly separates questions regarding what we can observe from questions of belief—what things are from *why* things are the way they are. Second, we see an explicit critique of religion characteristic of modernist atheists. The progressionist view—which, as Ruse put it, is "a matter of rolling our sleeves and doing it ourselves"—is viewed as superior to the providential view, in which contemporary problems are placed in God's hands. Third, the narrative places a premium on modern science—or more specifically cognitive rationality—as the primary avenue toward progress.

Morality Without Religion

In a research article published a few years back, psychologist Will Gervais explained that, "Scientific research yields inconsistent and contradictory evidence relating religion to moral judgments and outcomes, yet most people on earth nonetheless view belief in God (or gods) as central to morality, and many view atheists with suspicion and scorn. . . . American participants intuitively judged a wide variety of enormously immoral acts (e.g., serial murder, consensual incest, necrobestiality, cannibalism) as representative of atheists, but not of eleven other religious, ethnic, and cultural groups. Even atheist participants judged immoral acts as more representative of atheists than of other groups."[50] These findings have been replicated cross-nationally.[51]

And yet many atheist scientists we interviewed are frustrated by the perception that atheists are immoral and want to counter the idea that God is a requirement for living a moral life and doing the right thing. One graduate student in physics—[52]who describes himself as having shifted from being a "fire-breathing atheist" to a "certain atheist"—said:

I just believe in being good to people. And the thing that I find most frustrating is that there are some Christians in the country who think that atheists have no moral code, which I find fairly ironic since I believe in treating everyone equally, especially women and people who are not straight. . . . Why should we hate on them because a book tells us, you know?

Another graduate student in physics[53] explained to us:

It's not a God-ordained purpose . . . [but] I want to leave something for the future. . . . It's incompatible with my worldview to
lead a totally hedonistic lifestyle in which all you cared about was
yourself. I would find that wrong for ethical and moral reasons.

To explore the link between atheism and morality, we presented
scientists in the U.S. and U.K. with a series of statements tied to
moral values to assess those with which they identify.[54] One statement was designed to emphasize equality. It read: "I think it is important that every person in the world be treated equally. I believe
everyone should have equal opportunities in life." In our survey,
there was virtually no difference in how religious and atheist
scientists responded to this question. Across both groups, roughly
97 to 99 percent of scientists responded that this statement is
"like me."

We saw a similar pattern when we looked at another self-
transcendent value: empathy. During an interview with one atheist
graduate student in physics,[55] he referenced this virtue when we
were discussing what it means to be a "responsible scientist" and
stated, "It means being able to listen to everyone's story. . . . It means
being open and being able to understand where different groups
are coming from and not just dismissing people outright." To explore how atheist scientists approach empathy, we asked U.S. and
U.K. scientists whether they agreed with the statement, "It is important to me to listen to people who are different from me. Even
when I disagree with them, I still want to understand them." Once
again, there was virtually no difference between religious and
atheist scientists, with 97 to 98 percent of all scientists responding
affirmatively.

Then we looked at values related to power and the pursuit of
self-interest. The stereotype of both scientists, who work in a
competitive industry, and atheists—who celebrate natural evolution based on "survival of the fittest"—might lead us to believe
that atheist scientists would score much higher on these values

than religious scientists. In our survey, we found that religious and atheist scientists are different when it comes to the pursuit of power and self-interest, but not in the way we might think. Atheist scientists were slightly less likely than religious scientists to agree with the statement, "It is important to me to get respect from others. I want people to do what I say." Among scientists in the U.S. and U.K., 71 percent of religious scientists said this statement was "like me" compared with 67 percent of atheist scientists in the U.S. and 65 percent of atheist scientists in the U.K. Modernist atheists—those furthest removed from religion—were actually the *least* likely among the groups to espouse agreement with this notion of power. Between 62 and 63 percent of modernist atheist scientists in the U.S. and U.K. indicated agreement with the statement.

Likewise, when we examine values related to self-enhancement, we find that atheist scientists are substantially less likely than religious scientists to embrace values that emphasize the pursuit of self-interest and power over others. Existing research on atheists in the general population tends to reflect the patterns we observe here.[56] When asked about the importance assigned to being rich, for example, the gulf between atheist and religious scientists grows larger. Of religious scientists in the U.S. and U.K. 41 percent agreed with the statement "It is important to me to be rich. I want to have a lot of money and expensive things." Only 34 percent of atheist scientists in the U.S. and 29 percent of atheist scientists in the U.K. agreed with the statement.

Most of the atheists we spoke with explained that they see moral values such as equity and empathy as innate or autonomously defined, rather than dictated by God or informed by serving God. One modernist atheist scientist,[57] for example, told us:

> You don't really need a God to give you a moralistic framework. You experience the world and you understand the world with a set of morals.... If you have personal values and personal morals,

you should be able to delineate what's good and bad without a higher power telling you "Don't do that."

A culturally religious atheist[58] similarly emphasized a normative morality without God, saying:

I believe that there's an external set of moral facts that we can approach through thinking and reasoning. But it's very hard to justify philosophically why that should be true. So you have to sort of believe it's true. Like I believe that killing people is wrong, not because God tells me to, but because I believe there's an external moral factor that says that killing is wrong. But it's hard to really explain why that necessarily has to be true beyond the fact that humans agree that it is true.

Atheist scientists hope to dispel the idea that religion's loss results in a loss of values and morality. As they see it, morality is not imbued by God or religious belief, but rather arises from reason and a rational examination of how to improve well-being and make the world a better place. The atheist scientists we spoke with challenge the idea that religion is necessary to uphold values like equality and empathy, and there is some evidence that public opinion might be starting to change. While 42 percent of Americans in a 2017 Pew survey said belief in God is necessary to be moral, 56 percent said belief in God is not necessary to be moral and have good values, up from 49 percent in 2011.[59] Also, while many people still believe their morality comes from God or belief in God, an interesting experiment by Nicholas Epley and his colleagues shows that what many people think of as religious morality might not be what it seems. "Experimental evidence suggests that people's opinion of what God thinks is right and wrong tracks what they believe is right and wrong, not the other way around," writes Jim Davies, a professor at the Institute of Cognitive Science at Carleton University, in a recent article.[60] "Social psychologist Nicholas Epley and his colleagues surveyed religious believers about their moral beliefs and the moral beliefs of God. Not surprisingly, what people thought was right and wrong matched up pretty well with what they felt God's morality was like. Then Epley and his fellow researchers attempted to manipulate their participants' moral beliefs with persuasive essays. If convinced, their moral opinion should then be different from God's, right? Wrong. When respondents were asked again what God thought, people reported that God agreed with their new opinion!"

Some Transcendent Meaning? Atheists Who Believe in Something Greater

In Chapter 5, we met several spiritual atheists at the end of the chapter that *did* feel as though their life had meaning, purpose, and drive, not only in this life, but even potentially in the next. We remind our readers, in particular, of the U.S. biology professor[61] who actually connected her science, atheism, and spirituality to a conviction that she would one day reunite with her late loved ones:

> So if you think about what happens when you die, I think one of the things that's been most meaningful to me as a science discovery is that all the bits and pieces of us, carbon, atoms, etc., were once out in the stars . . . so if you do calculations as to kind of how you've been built, where those elements come from and where they're going, it's a much more cosmic picture than I might have imagined. And that's been particularly helpful for me when I think about, for example, the death of my grandmother who was not religious, and thinking about "when will we ever be together again?"

A nihilistic atheist, at this point, would be tempted to answer, "of course not!" If this life is all there is, then once someone is gone, they are gone for good. There is no hope, no purpose, and no meaning;

nothing to generate guidance that would lead to a reunion. This atheist, however, does not respond this way, "and I think, well, we will be together again. At some point the carbon in me will [integrate] and we'll be together again!"

We return to this quote as a reminder that atheism is not monolithic—we have seen over and again that atheist scientists shatter our expectations. Some atheist scientists are indeed pessimistic and nihilistic, without hope and with a sense of meaninglessness and lack of purpose. Many however, find purpose in their craft, in the progress they and their colleagues make, all while admitting that meaning *might* exist, we just cannot see it from our vantage point. Still others *do* find meaning and purpose in life, be it from their families, meditation, or the conviction that their atoms will once more dance with those of their loved ones through the expanses of the universe.

8

From Rhetoric to Reality

Why Religious Believers Should Give Atheist Scientists a Chance

Many religious people believe all scientists are atheists and all atheist scientists are like Richard Dawkins. In other words, for religious individuals, the New Atheists and their positions—hostile to religion, derisive, using acerbic language to attack religious faith—are often viewed as representative of scientists and science. This leads most religious individuals to believe that most atheist scientists share the goal of eliminating religion. Psychologist Elizabeth Barnes, for example, found[1] that Christian students at secular universities believe scientists harbor greater bias against them than these scientists actually do, which likely impacts how much these students trust scientists and how well they do in science courses. Such views may make religious individuals reluctant to send their children to secular universities and, in particular, to major in scientific fields. The New Atheist monopoly on the image of the atheist scientist is a marketing problem for the scientific community on multiple levels.

Science Has a Marketing Problem

We believe the rhetoric of the New Atheists has created problems for the scientific community by weaponizing science against religion—yet, our research shows that the *rhetoric* is very different from the *reality* of atheism in science. The reality is this: there are

Varieties of Atheism in Science. Elaine Howard Ecklund and David R. Johnson, Oxford University Press.
© Oxford University Press 2021. DOI: 10.1093/oso/9780197539163.003.0008

different kinds of atheists in science and they are not all like the New Atheists. As our research shows, most atheist scientists in the U.S. and U.K. are not anti-religious. Most, in fact, think the tone of the New Atheists is problematic. While atheist scientists tend to view science as the best way to gain knowledge and discover truths about the natural world, they do not view it as the only way to understand the world around us. Some even think religion can be beneficial to society and disagree with the condemning way New Atheists frame public discussions of religion. Many atheist scientists still find cultural aspects of religion important, actively engaging in religious congregations, schools, and communities. To our religious readers, we say this: our sociological research shows that most atheist scientists do *not* want to eliminate religion from society or destroy your faith community.

We think it is incredibly important for religious believers to know what atheist scientists truly think about religion. In the U.S. and the U.K., countries at the core of the global science infrastructure, public confidence in science has consistently remained high since survey researchers began tracking it in the early 1970s.[2] In recent years, however, some members of the scientific community have grown increasingly concerned about the erosion of trust between scientists and the public, especially segments of the religious community. In the U.S., nearly one in four people believe that scientists are hostile to religion.[3] Concerns about hostility are especially pronounced among a subset of religious Americans (particularly evangelical Christians). Roughly two-thirds of these individuals believe most scientists are hostile to religion.[4] Muslims in the U.K. have also expressed concern about science and the attitudes of outspoken atheist scientists toward Islam.[5]

For some observers, the solution for lack of trust in science is to improve public understanding of science. This is what is known as the "deficit model" of science communication—the idea that skepticism toward science is the result of a lack of scientific knowledge, and it will help if people are simply given more information. A large body of research has discredited the deficit model, however.[6] What is more, sociological research consistently shows that most religious Americans are no different from nonreligious Americans in their knowledge of science or interest in science; this is true even for evangelical Christians, a group that has been pitted again the science community in the public sphere.[7] "Instead of simply increasing public understanding of science, scientists need to have real dialogue with members of the public, listening to their concerns, their priorities, and the questions they would like us to answer," argues Alan Leshner, former CEO of the American Association for the Advancement of Science (AAAS).[8] Unfortunately, when scientists do engage the public, they often do not seek out the groups for which dialogue may be most necessary. Scientists need to communicate beyond the "proverbial choir," notes Dietram Scheufele, a leading expert in science communication at the University of Wisconsin–Madison. In his view, "the need to reach segments that we have traditionally not connected with as effectively as we should have is more urgent than ever before."[9] We think scientists need to be better about reaching out to religious individuals and communities—and religious individuals and communities need to be more receptive to scientists. We also think that religious believers will be more receptive to scientists if they have a better understanding of scientists—especially atheist scientists.[10]

Toward a New Public Understanding of Atheist Scientists

What do religious individuals need to know about atheist scientists? Our findings suggest four messages. The first is that many atheist scientists in the U.S. and U.K. are not hostile to religion. We did find some anti-religious sentiment among atheist scientists in our study, particularly among modernist atheists, whose lives are generally segregated from religious individuals and organizations. In our interviews with these scientists, some did voice concerns about the role of religion in society, particularly its impact on cognitive rationality. Our survey data show that nearly two-thirds of these scientists in the U.S. and U.K. do believe science and religion are in conflict with one another, but this does not necessarily connote *hostility*. We found many other modernist atheist scientists who see religion playing a positive role in society and reject anti-religious sentiment from scientists in the public sphere. Culturally religious atheists choose to embrace elements of religion in their day-to-day lives and intimate relationships. Those who were raised in minority religions especially often still view their religious identities as important and continue to identify with their childhood traditions. They see value in being a part of a religious community and often engage in religious practices. Though spiritual atheists do not partake in the traditions and practices of organized religion, we cannot conclude that most spiritual atheists are anti-religious.

Second, atheist scientists are not atheists for the reason many religious believers might think. Science is often integral to the atheist identity, but not necessarily the cause of it. Our work reveals that scientists are atheists for a multitude of reasons, and exposure to science proved to be less important in developing their nonbelief than did other factors. Most atheist scientists were raised in irreligious households, so they were primed to embrace atheism well before they started their scientific training. Atheist scientists who have turned away from religion often did so because of experiences within their childhood religious communities, usually before their scientific training began. Others fell into social networks that led them to question their religion before ultimately giving it up. While in both the U.S. and U.K. roughly half of atheist scientists indicated on surveys that scientific knowledge made them less religious, we found that scientific training was rarely the reason these scientists embraced atheism. Interviews with atheist scientists who took our survey suggest that exposure to science affirmed prior doubts about religion or accelerated trajectories away from it rather than catalyzing these ideas and pathways. These findings should help allay concerns religious individuals have about their children attending postsecondary institutions where science is emphasized or pursuing careers in STEM. Our findings reveal that scientific education and knowledge will not necessarily lead them away from their faith.

Third, not all atheist scientists embrace scientism. Scientism demands that all knowledge, all ways of seeing and understanding the universe, must reduce down to science, and any other generative worldview is silly at best and dangerous at worst. Most atheist scientists eschew the hubris of scientism. They are fierce proponents of science and exude confidence in its intrinsic value and power to understand and transform the world, but this is science advocacy, not scientism. We did find a minority of modernist atheist scientists who do embrace the most intense form of this view, asserting that science—by which they mean natural science—is the only legitimate way to understand the world, but most atheist scientists we met did not present science as the only source of knowledge. While these atheist scientists believe science is the most important tool for explaining and understanding the natural world, and often anchor meaning and morality around scientific achievements and advancements, they believe that science does have limits and boundaries. Many also recognize a role for other bodies of knowledge and ways of knowing, including religion, philosophy, and ethics.

Finally, atheist scientists are not as dissimilar from persons of faith as the rhetoric of New Atheists suggests. In fact, they share many of the same feelings and values, even if the root of their core commitments is different. Many atheist scientists are devoted to enhancing human well-being and the world around us, just as religious individuals are dedicated to the same. They experience awe, wonder, and humility and, as we've seen, a number of them embrace spirituality and religious culture more specifically.

Pathways of Dialogue

How, then, can individuals and organizations align the rhetoric and reality of atheism in science, and who should be involved? We suggest greater dialogue between atheist scientists and religious communities, something that rarely happens currently. Social scientists have found that prejudice and erroneous assumptions are reduced when members of different groups—including ones historically in tension with one another—come together in meaningful ways and emphasize the common aspects of their identities.[11]

At the national level, major scientific organizations, such as associations for scientific disciplines and societies, could establish programs or working groups that seek to facilitate communication between scientific and religious communities. An exemplary model is found in the American Association for the Advancement of Science's Dialogue on Science, Ethics and Religion (DoSER) program,[12] a 25-year-old initiative that uses workshops, symposia, lectures, and other events to help scientists promote constructive dialogue on science and religion. The central participants in such programs are often religious scientists and faith leaders; one of the key opportunities for growth would be more involvement of nonreligious and atheist scientists as well as religious congregants. The association is establishing collaborative relationships between national scientific and denominational organizations and developing programming that targets religious communities, leaders, and faculty in seminaries.

At the local level, we think efforts should emphasize creating informal, in-person, or online interactions where atheist scientists and religious believers can get to know each other. We think this would show religious individuals that most atheist scientists do not carry themselves the way New Atheists do on Twitter. For these events, we don't think organizers necessarily need to be concerned with including religious scientists, or centering dialogue on topics that have an apparent connection to faith, but rather they should focus on providing a space where atheist scientists and religious believers can find their own connections. While there is a place for events that actively target atheist scientists in discussion and debate about science and faith, for conferences that examine the tension between science and religion, and for seminars that delve into the connection between the two, we don't think this is where the majority of such contact should occur; research shows that when differences and boundaries between groups are highlighted, stereotypes are reified, not rectified.

Further, university science communicators could run events at pubs and coffeehouses that are marketed to local religious groups. Religious student organizations on campus could be helpful in this regard, connecting these communicators with local faith communities. By reaching out to religious congregations specifically, the scientific community sends a message that it is open to dialogue with religious believers. BioLogos, a Christian organization that has sought to convince other Christians of the veracity of evolution, is doing this kind of work for Evangelicals by sending scientists to congregations to talk about their scientific work.[13] As of 2018, they've reached nearly 11,000 people through these and other events.[14]

For their part, congregations, faith leaders, and youth group ministers can reach out to local scientists, inviting them to come talk about their work, without concern for their religiosity. Our data suggest that only a subset of modernist atheist scientists would have reservations about such an invitation and that most other scientists—atheists, agnostics, or otherwise—would likely accept the opportunity to discuss their work with communities of faith. One approach would be to tie such events to the religious calendar. During Ramadan, for example, a mosque could invite in a biologist to discuss what happens to our bodies and minds when we fast, or Earth Day could serve as a good opportunity to organize lectures by environmental scientists. While groups like BioLogos, Science for the Church, and Sinai and Synapses[15] are already running these sorts of discussions with congregations, these groups more often focus on having religious scientists in dialogue with religious congregants. They rarely bring in atheist scientists as speakers or participants, which does nothing to dispel the myth that atheist scientists are anti-religious.

Why We Should Care

We don't want to pretend that such efforts will not be challenging. Ideological segregation and political polarization tend to color how we understand one another, after all. But changing how religious people view the scientific community, specifically the atheists within it, is too important not to try.

The scientific community should be concerned about its relationship with the religious community for a couple of related reasons. First, public trust and confidence in science is important because it has implications for research and development funding, the vitality of universities and research institutes, the prestige of the science profession, and how individuals process information about science. At the time of this writing, religious views about science are having an impact on how we respond to COVID-19, the teaching of evolution, the acceptance of vaccines, anti–climate change efforts, to name just a few important issues.

Second, how religious communities perceive the scientific community has consequences for diversity in science. Nearly every major scientific organization in the U.S. and U.K. is working toward, and spends considerable resources on, attracting and increasing the participation of underrepresented groups in the scientific workforce. As the U.K.'s Royal Society, the oldest independent scientific academy, explains in its Diversity Strategy for 2019–2022, "a diverse and inclusive scientific community that brings together the widest range of talents, backgrounds, perspectives and experiences, maximises scientific innovation and creativity, as well as the competitiveness of the U.K. scientific industry."[16] Religion is an integral dimension of these goals, but rarely recognized as such.

In the U.S. and the U.K., women and non-white individuals—the core groups targeted by diversity initiatives—are underrepresented in science, but they are overrepresented in religious traditions. In other words, women and minorities are more likely than other groups to identify as religious.[17] Thus, if the scientific community wishes to identify and address barriers to participation in STEM and increase diversity, they need to address their relationship with religious communities, which currently often see science as a profession that is hostile to their values and beliefs.

Just as the scientific community would benefit from a better relationship with communities of faith, the religious community stands to benefit from dialogue with the scientific community. In the same respect that scientists should not be stereotyped based on a vocal minority of atheists who espouse anti-religious views, the religious community should not be judged according to stereotypes. But if these two groups continue to eschew one another—or worse, embrace conflict over dialogue—persons of faith will remain misunderstood by scientists. In addition, religious people care about truth. And even if there are instances where these two communities disagree about truth claims, religious people should want to believe truth, not stereotypes, about atheist scientists and science. If religious people really believe that truth resides in God, then the pursuit of truth in science can *only* lead them closer to God.

Even as atheists remain marginalized in many ways, they also represent a visible—and by some measures, growing—community. Given the centrality of science to what it means to be an atheist, religious communities should always embrace dialogue—even with those with whom they disagree. Having a fuller understanding of both religious and atheist scientists will help religious communities better foster interest in science among those in their congregations (particularly children) and be better conduits for science justice,[18] encouraging more women and underrepresented racial minority groups to enter science.

For too long, the New Atheists, and their hostility to religion, have erroneously been viewed as representative of atheist scientists, and often scientists more broadly, in the U.S. and U.K. In large part, this is because of the bestselling books of Richard Dawkins, Dan Dennett, Sam Harris, and Christopher Hitchens—the "Four Horsemen" of New Atheism—who created a publishing and social media phenomenon that led to a de facto monopoly over how the public understands atheist thought and atheism in science. Their work and words are seen as assaults on religious faith. But the atheists we found in the science community were, for the most part, very different. Atheism in science is more complex, more nuanced, and more accepting of religion and spirituality than the New Atheists have led us to believe, and it is now our responsibility to replace their rhetoric with reality.

APPENDIX

Studying Atheist Scientists

There are at least two ways to approach social science research. In one scenario, researchers enter the field with a specific question and a series of expectations or hypotheses about what they will find. In another scenario, researchers similarly have a question that leads them into the field, only to discover that a different question begs their attention. The former approach is often associated with survey research and held up as the ideal—and, given the need to have standardized instruments, it is often impossible to chase a different question once fieldwork begins. The latter approach is often associated with ethnographic research, in which researchers situated within a social setting conduct interviews and observations. The pace and epistemic nature of ethnography means researchers are able to pursue different questions relatively easily after commencing a project. This book falls somewhere between these two approaches.

The data we based this book on came from the Religion Among Scientists in International Context Study (RASIC). RASIC examined how scientists, specifically physicists and biologists, in eight different countries and regions view the relationship between religion and science. The study was a mixed-methods effort of massive scale that included a survey of 22,525 scientists in the United States, the United Kingdom, France, Italy, Turkey, India, Hong Kong, and Taiwan and 609 in-depth interviews with selected survey participants. We conducted pilot interviews during 2011 and 2012. The survey took place between 2013 and 2015 and received 9,422 responses. Region-specific response rates ranged from 39 percent (Turkey) to 57 percent (Italy and the United States), with an overall response rate of 42 percent. While a number of countries represent interesting contexts in which to study scientists' views of science and religion, the sampling design emphasized variation in religiosity of the general population, science infrastructure, and state policies toward religion.

One of the key goals of RASIC was to understand how scientists respond to religion. Naturally, we already knew that secularism is integral to understanding religion in science and we designed all of our questions on the survey and interview instruments such that anyone could answer them independent of their religious beliefs or lack thereof. Employing the survey research approach, we entered the field with key questions about scientists' *religious* identities, practices, and beliefs, and our answers to those questions are reported in our book *Secularity and Science: What Scientists Around the World Really Think about Religion*, as well as a series of social scientific articles.[1]

Nevertheless, much like ethnographers discovering a different question in the midst of fieldwork, the importance of better understanding *atheist* identities in science slowly percolated over time and became particularly apparent once we made significant progress analyzing data for *Secularity and Science*. If researchers were watercraft, ethnographers would be windsailers—highly nimble solo-operators who can change direction (with few moving parts) based on information they are collecting around them, such as changes in the wind direction. International comparative projects such as ours are more like cruise lines—they require a large team, there are several moving parts that impact one another, and they need to arrive at preplanned destinations at preplanned times. This means that changing direction is somewhat more complicated and not always desirable. It wasn't until we had finished the first book from the RASIC data that we recognized how unique, interesting, *and varied* the atheist scientists were and decided to dive deeper into atheism.

Yet, looking back, the first clue that it was important to better understand atheist scientists, and the variation among them, was revealed in *the very first* interview we conducted in the U.K. In the field notes we write after each interview, we describe the setting, respondent, and various elements not recorded on the interview transcript. In the notes from that first interview, with an atheist scientist,[2] there is evidence that this category of scientist is distinctive and misunderstood. We wrote:

> Several times he mentioned what an important study it was, pointing out that there are a lot of misconceptions about what British scientists think about these topics, especially because of Richard Dawkins' work and that rarely does he think scientists are at all like Dawkins.

This scientist's point represents one of the empirical endeavors we undertake in this book—examining to what extent New Atheist scientists like Richard Dawkins and others are representative of the scientific community more broadly.

The interviews in the U.K. represented our research team's first study of science and religion abroad, and Elaine, with one study of religion among scientists in the U.S. under her belt,[3] noted after an interview with an atheist scientist[4] in Britain:

> I keep noticing how free I feel with academics in Oxford and Cambridge compared to those I have interviewed about similar topics in comparable U.S. universities. . . . Frankly, the scientists here, although less religious on the whole, are just plain friendlier and easier to talk with. It's striking actually.

To be sure, some scientists we met with seemed suspicious about our interest in surveying and interviewing them about religion in science. We ultimately found that U.S. and U.K. atheist scientists are not vastly different from each other, but one important difference could be their orientations to conversations about religion and science. In the U.S., the conflict narrative is much more pronounced in the public sphere than in the U.K. This difference could lead U.S. scientists to be somewhat more reticent than U.K. scientists to discuss religion.

Our qualitative work also pointed to notions of atheism and identity. For one, from time to time we found trappings of atheist symbolism in the scientific workplace. In one department of physics at a university in Maryland, for example, we noted that one scientist had posted a newspaper article entitled "Why God Did Not Create the Universe" on his office door. We also observed New Atheist texts displayed in scientists' offices; we noticed many copies of Richard Dawkins's *The God Delusion,* for example. According to our context notes from another interview in the U.K.,[5] a biologist told us in advance of his interview that he is "a strong atheist" and that he's "read Dawkins, Sam Harris, Hitchens, and un-named authors from the other perspective who make the case that religion and science are not at war with one another." Ultimately, we started asking scientists about books they've read related to religion and science. After we completed data collection in the U.K., our preliminary analyses also pointed to a tendency among scientists to bring up Richard Dawkins in their interviews.

Perhaps the most persistent notion of atheist identity was the tendency to announce it at the outset of an interview, before any question about religion even came up. While we reserved questions about the scientist's religious views until mid-interview, we grew accustomed to atheist scientists beginning interviews like this:

> I guess the best thing for me to start just to say outright is I'm an atheist.[6]
>
> OK, from my personal point of view religion doesn't [factor] into it at all. I'm an atheist.[7]
>
> To me personally, I'm pretty much a complete atheist.[8]

Subsequently, we began to wonder about varieties of atheism in science and turned to all of our RASIC data to determine what we could learn. On the one hand, had we possessed the foresight to recognize the importance of social differentiation and identity among atheists in science, we would have created survey and interview instruments that could have generated even more data than what we present in this book. As we note at the outset of the book, we did not assume the differences among atheists would be so stark. On the other hand, however, we did spend hundreds of hours talking about religion and belief with atheist scientists and are able to use our survey data to look at a number of important dimensions of social life among atheist scientists. It is in this respect that we entered the field with one question, but are nevertheless able to address a different and equally compelling question after we launched RASIC.

Given that the data for this book were generated by RASIC, the remainder of the appendix provides the essential methodological details needed to understand the origins of this book.

Sample Procedures at a Glance

We detail the full scope of the RASIC sampling procedure in *Secularity and Science*. Here, we provide selected details that are essential to understanding how we surveyed scientists in the U.S. and U.K. The sampling process entailed two stages that occurred between 2011 and 2012, with stage one involving the selection of organizations and stage two involving the selection of individual scientists.

The organizational context of science is important because the structure and culture of scientific work varies widely within and across regions of the globe. Differences in financial and physical resources, prestige, norms and values, collaborators, and standards of achievement, for example, generate differences in conditions of scientific practice.[9] In principle, differences in organizational context may also be related to religion and nonreligion—if, say, nonreligious scientists are more likely to be found in elite versus non-elite organizations. Accordingly, stage one of the sampling process entailed the construction of a sampling frame of organizations (universities and institutes) that we stratified by discipline (physics and biology) and status (elite and non-elite). Within biology, we focused on core subfields, including cell biology, developmental biology, structural biology, molecular biology, biochemistry, neuroscience, immunology, microbiology, genetics, plant science/botany, animal-related research, zoology, physiology, nutrition, ecology, evolution, infectious disease, and other selected medical subfields. We excluded more interdisciplinary subfields, such as computational biology and cognitive science, because their categorization would not fit easily within our schema. Physics departments were much less varied and thus we mainly excluded subfields such as engineering and earth science.

The main challenge in sampling science organizations around the world is that there is not an exhaustive list of universities or, for that matter, physics and biology programs in a region. Even where lists are present, such as those generated by the National Research Council in the U.S. or the Research Excellence Framework in the U.K., they are generated according to different evaluative criteria, meaning that such lists produce different types of programs and employ different rankings. To identify the universe of physics and biology organizations, therefore, we examined the institutional affiliations of authors on articles published between 2001 and 2011 in the Thomson Reuter Web of Science database. The Web of Science includes more than 12,000 scientific journals worldwide, primarily in English—the main language in which scientific articles are written—but also includes some regional journals in local languages. In each region, we focused on publications in physics and biology and generated a list of up to 500 organizations with which authors are affiliated and ranked the list according to articles per institution.

Because institutions vary widely in research productivity, we stratified institutions by research productivity in two categories, "elite" and "non-elite." This strategy ensured that we included fewer research-oriented universities. To do so, we employed a triangulation process comprising research productivity (the number of times an organization appeared as the affiliation of an author), insider opinions (evaluations by scientists in each region of the country), and in-country ranking systems. From this stratified sampling frame, we then selected enough elite and non-elite organizations to generate 2,000 scientists per discipline in each region. Overall, this included 662 organizations across the eight regions of the study: 102 elite biology organizations, 146 non-elite biology organizations, 112 elite physics organizations, and 220 non-elite physics organizations.

In each organization, we used departmental websites to construct the sampling frame of individuals. We stratified this sampling frame by gender and career stage. Ranks vary across national contexts, both in terms of titles used and what they signify. Accordingly, we created three categories for rank. Rank 1 encapsulated scientists in training (graduate students). Rank 2 captured junior scientists who have finished their training and work as postdoctoral fellows, early-career instructors, or tenure-track faculty members. In the U.S., this included postdoctoral fellows, assistant professors, research professors, and research scientists. In the U.K., Rank 2 included postdoctoral fellows, assistant professors, lecturers, readers, and research fellows/associates/assistants and academic staff with a Ph.D. Rank 3 encompassed advanced scientists. In the U.S., this included associate professors and professors. In the U.K., Rank 3 included associate professors, full professors, Royal Society Fellows, "Named" Fellows, research fellows with prestigious grants, and "statutory professors."

In each discipline in each region, this resulted in a sampling frame with 12 strata in terms of institutional type, gender, and career stage: *2 institutional statuses (elite/non-elite) x 2 genders (male/female) x 3 academic ranks = 12.* We ultimately attempted to compile a sampling frame of at least 2,000 scientists for each discipline in each nation to ensure statistical power in each stratum. In each region, respondents were given the option to take the survey online or by phone. An overwhelming majority of respondents completed the survey online. Surveys were offered in the local language of each region and English. In the U.S., we provided a sampling file with 3,989 names to the survey firm

Abt SRBI, who fielded the survey between January 14 and March 23, 2015. Each individual contacted received a $5 preincentive to participate. A total of 1,989 scientists completed the survey, generating an adjusted response rate of 57.1 percent.

In the U.K., the first national context in which we conducted the study, the sampling process differed slightly from other regions. Rather than focusing on organizations (e.g., universities), we began the study by generating a sampling frame of suborganizations (departments). University prestige and departmental prestige are often correlated, but do not always align. This initial sampling strategy was motivated by a desire to be thorough and the Research Excellence Framework in the U.K. allowed us to examine discipline-specific rankings of each department. However, as we proceeded to subsequent regions, we discovered such discipline-specific rankings were not consistently available and thus we embraced university prestige (alongside insider ratings by in-country scientists and publication counts) as a reliable proxy for departmental prestige. A second way in which the development of the sampling frame differed in the U.K. was stratification by region (England, Northern Ireland, Scotland, and Wales). This generated 16 first-stage strata: *2 disciplines (biology/physics) x 2 statuses (elite/non-elite) x 4 regions = 16.* We attempted to sample departments equally by discipline and status and in proportion to the total population of the region. This approach generated a sampling frame of 3,393 scientists in the U.K. The U.K. data collection firm GfK NOP administered the survey from September 19 to October 16, 2013. All scientists in the sampling frame were provided a £5 preincentive. A total of 1,604 scientists completed the survey. After postsurvey adjustments, we achieved a response rate of 50 percent.

Qualitative Interview Sampling

In all regions, our survey instrument included a question asking respondents whether members of the research team could contact them for a follow-up interview. The sampling frame of interview respondents therefore consisted of any respondent who answered this question affirmatively. Because of the centrality of religion to the study, we then stratified the interview sampling frame according to reported religious identity by using the following question prompt: "Independent of whether you attend religious services or not, would you say you are . . ." Response choices to this question included: 1) "A very religious person"; 2) "a moderately religious person"; 3) "a slightly religious person"; 4) "not a religious person"; 5) "a convinced atheist"; and 6) "don't know." We collapsed respondents into three categories: "religious" (1 and 2), "slightly religious" (3), and "nonreligious" (4 and 5). This approach ensured that our interview sample would represent the broadest range of religious views. In each region, we sought to balance the interview sample across these three categories. Within each category and in each region, we sought balance on other dimensions of the study such as discipline, gender, rank, and institutional status. In all regions of the study, we offered to conduct interviews in the local language or English.

In the U.S., 559 of the 1,989 scientists who completed the survey agreed to be contacted for a follow-up interview. We interviewed 100 scientists in the U.S., with 55 in-person interviews conducted in North Carolina, Maryland, Texas, California, and New York. To reach scientists in other regions of the U.S., we conducted 38 interviews using the telephone and seven using Skype. Thirteen of these 100 interviews were conducted during the pilot stage, prior to fielding the survey.

In the U.K., 643 of the 1,604 survey respondents agreed to be contacted for a follow-up interview. Prior to fielding the survey, our team conducted 39 pilot interviews. We interviewed 98 survey respondents, for a total of 137 interviews in the U.K. Given the number of willing interview participants, we were able to balance the interview sample across all of the key dimensions of the study. Most of the interviews (89) were conducted in-person, primarily in Oxford, Cambridge, London, and the West and East Midlands region of England (Coventry, Birmingham, and Leicester). The remaining interviews were conducted by telephone (33) and Skype (15) in other regions of the U.K.

How We Identified Modernist Atheists, Spiritual Atheists, and Culturally Religious Atheists

To begin this project, we needed to delimit the U.S. and U.K. RASIC data to atheists. This alone was easy given that we asked about belief in God on the survey. After a preliminary analysis of interviews with atheists, however, we needed to identify modernist, spiritual, and culturally religious atheist scientists in the survey data. To do so, we used a series of questions from the RASIC survey instrument. These included questions about belief in God, frequency of prayer, attendance at religious services, spirituality, and religious affiliation of spouse. The questions appeared on the RASIC survey as follows:

GOD. Please indicate which statement below comes closest to expressing what you believe about God. Would you say ...?

<1> I don't believe in God.

<2> I don't know whether there is a God and I don't believe there is any way to find out.

<3> I don't believe in a personal God, but I do believe in a Higher Power of some kind.

<4> I find myself believing in God some of the time, but not at others.

<5> While I have doubts, I feel that I do believe in God.

<6> I know God really exists and I have no doubts about it.

PRAY. Now thinking about the present, about how often do you pray?

<1> Never

<2> Less than once a year

<3> About once or twice a year

<4> Several times a year

<5> About once a month

<6> 2–3 times a month

<7> Nearly every week

<8> Every week

<9> Several times a week

<10> Once a day

<11> Several times a day

ATTEND. Apart from weddings and funerals, about how often do you attend religious services these days?

<1> More than once a week

<2> Once a week

<3> Once a month

<4> Only on special holy days

<5> Once a year

<6> Less often

<7> Never, practically never

SPRTLTY. What best describes you?

<1> I follow a religion and consider myself to be a spiritual person interested in the sacred and the supernatural.

<2> I follow a religion, but don't consider myself to be a spiritual person interested in the sacred and the supernatural.

<3> I don't follow a religion, but consider myself to be a spiritual person interested in the sacred and the supernatural.

<4> I don't follow a religion and don't consider myself to be a spiritual person interested in the sacred and the supernatural.

RLGSPAFF1. Does your spouse or partner belong to a religion or religious denomination? If yes, which one?

<0> I do not belong to a religion

<1> Roman Catholic

<2> Protestant

<3> Orthodox (Russian/Greek/etc.)

<4> Jew

<5> Muslim

<6> Hindu

<7> Buddhist

<99> Other [specify] {RLGSPAFF1_TXT}[10]

The overall subset of atheists in the U.S. and U.K. was selected by identifying survey participants who indicated that they do not believe in God. We categorized modernist atheists as atheists who do not believe in God, never attend religious services, never pray, do not indicate interest in spirituality, and who are not in a relationship (partner or marriage) with a religiously affiliated individual. Spiritual atheists meet the same conditions as modernist atheists, save for agreeing with the statement on the spirituality question: "I don't follow a religion, but consider myself to be a spiritual person interested in the sacred and the supernatural." Culturally religious atheists are atheists who do not believe in God, but meet one of the following conditions: pray at least some of the time, attend religious services at least some of the time, or follow a religion but are not interested in the sacred and the supernatural. Roughly 2.5 percent of the atheist scientists in the U.S. and U.K. could not be categorized in this typology because of missing values (e.g., they did not answer a question about prayer). We excluded such cases as missing from analysis.

There were two circumstances in which we revised this classification in the qualitative analysis. First, in some interviews, we determined that atheists enrolled their children in religious primary or secondary schools, or had sustained interactions with religious organizations or communities. Because such information met our conceptual schema for being classified as a culturally religious atheist, we recoded them as such for the qualitative analysis. Second, none of the scientists who completed pilot interviews participated in the survey data collection process. As a result, a team of researchers read each pilot interview—which included many questions about religious beliefs, identity, and practices—and identified whether a given pilot interview participant was a modernist, spiritual, or culturally religious atheist.

Interviewing Atheist Scientists

After categorizing study participants as modernist, spiritual, or culturally religious atheists using the survey data, we then delimited our qualitative sample to the atheist scientists in each category. The RASIC interview guide broadly examined themes related to religion, ethics, gender, and work and family. It was organized in 10 sections, including: Academic and Research Background; Religion and Spirituality in Scientists' Work; Definitions of Religion, Spirituality and Science; Personal Relationship between Religion and Science; Religious History; Current Religious Identity, Beliefs, and Practice; Science Policy and Religion; Ethics in the Workplace; Research Integrity; and Women, Family Life, and Science. Overall, the interview guide included 38 questions.

For this book, three sections of the interview guide in particular generated most of the data that we analyzed:

Definitions of Religion, Spirituality, and Science

Now I'd like to ask you about your own understanding of religion and spirituality.

7. Could you give me your working definition of what those terms mean? What is your working definition of "religion," for example?
8. How about spirituality? Do you have a working definition of the term "spirituality"?
   a. *Probe*: Do you see something distinct from religion?
9. And while it may seem obvious, could you in just a few sentences give me your definition of science?
   a. [*If R doesn't understand*: I am interested in your own understanding of science. Can you tell me what that term means to you?]
10. Do you think there are limits to what science can explain?
    a. *Probe*: How did you come to this position?

Religious History

I also have just a few questions about your own religious history.

16. In what ways was religion a part of your life as a child?
    a. *Probe*: In what ways was religion talked about in your family setting?
17. [*Can skip, if redundant.*] Thinking about the arc of your life so far, has there been a time when you experienced a religious shift? Please tell me about that.
    a. *Probe*: It could be a small or large shift from religious to nonreligious, or nonreligious to religious, do you recall ever experiencing such a change?
    b. *Probe*: How about the relationship between religion and science, has there been a shift in how you view the relationship between religion and science?

Current Religious Identity, Beliefs, and Practice

18. How about now for you personally, how would you describe the place of religion or spirituality in your life? [*Or, if they have already talked about the place of religion or spirituality in their lives, ask*: Do you have anything more to add about the place of religion or spirituality in your life now?]
19. What religious or spiritual beliefs do you hold? How about religious practices?
20. In what ways, if anything, is religion a part of your family life now? [*Ask this question even if the respondent is not personally religious.*]
21. How do you answer the big questions of the meaning of life, such as why we are here, what is the meaning of my life? How do you find purpose?
22. Do you believe in God?
    a. *Probe*: Can you tell me a little bit about how you think about God or the concept of a God?
23. [*If no religious beliefs*] Do you believe in anything that is not readily observable?
    a. [*If R doesn't understand, restate*]: Do you have any beliefs about nonmaterial aspects of the world?
24. How about awe and beauty. Can you give me an example of any moments in your life when you have experienced a sense of awe, beauty, or wonder?
    a. *Probe*: How about in your scientific work. Are there occasions in your work that provoke these emotions?

Selected questions from other sections of the interview guide also proved relevant to our analysis, such as asking about views of the relationship between science and religion and religion in the workplace. For example, atheists' discussions of religious colleagues provide a window into whether they view religion in positive, negative, or neutral terms. And as we note earlier, many atheist scientists announced their atheist identity early in the interview, before we asked about their own religiosity or approach to religion.

Accordingly, our first step in the analysis was to read within each atheist category and analyze individual interviews in their entirety. This endeavor largely helped us understand the worldviews of the different categories of atheists and was central to writing Chapters 3, 4, and 5 in which we respectively describe modernist, culturally religious, and spiritual atheists. The interview section on current religious identity, beliefs, and practices generated a fair amount of the data in these chapters, given the nature of the questions asked. We then turned to other research questions and analyzed specific sections of the interview. Chapter 2 draws upon the religious history section to understand why atheist scientists transition away from religion (if religion was a part of their early life). Chapter 6 examines how atheist scientists define science to understand whether they believe there are limits to science. Chapter 7, on meaning and morality, largely draws from an interview question on meaning and purpose from the section on current religious beliefs. In each of these analyses, we analyzed interview responses across each atheist identity to determine how, if at all, perceptions vary among and between modernist, culturally religious, and spiritual atheist scientists.

Notes

Chapter 1

1. Egdell, Penny, Joseph Gerteis, and Douglas Hartmann. 2006. "Atheists as 'Other': Moral Boundaries and Cultural Membership in American Society." *American Sociological Review* 72(2):211–234; Ecklund, Elaine Howard, and Kristen Schultz Lee. 2011. "Atheists and Agnostics Negotiate Religion and Family." *Journal for the Scientific Study of Religion* 50(4):728–743.
2. Schmidt, Leigh Eric. 2016. *Village Atheists: How American Unbelievers Made Their Way in A Godly Nation.* Princeton, NJ: Princeton University Press.
3. Browne, Janet. 2003. "Charles Darwin as a Celebrity." *Science in Context* 16:175–194; Daum, Andreas W. 2009. "Varieties of Popular Science and the Transformations of Public Knowledge: Some Historical Reflections." *Isis* 100:319–332; Fahy, Declan. 2015. *The New Celebrity Scientists: Out of the Lab and into the Limelight.* Lanham, MD: Rowman & Littlefield; Giberson, K., and Artigas, M. 2007. *Oracles of Science: Celebrity Scientists versus God and Religion.* New York: Oxford University Press.
4. Ecklund, Elaine Howard, and Christopher P. Scheitle. 2017. *Religion vs. Science: What Religious People Really Think.* New York: Oxford University Press.
5. https://twitter.com/richarddawkins/status/507092728409522176; See, in particular, Hitchens, Christopher, Richard Dawkins, Sam Harris, and Daniel Dennett (foreword by Stephen Fry). 2019. *The Four Horsemen: The Conversation That Sparked an Atheist Revolution.* New York: Random House.
6. Kettell, Steve. 2016. "What's Really New about New Atheism." *Palgrave Communications* 2:16099.
7. Dawkins, Richard. 1992. "A Scientist's Case Against God." Presented at Edinburgh International Science Festival, April 15. Edinburgh, UK.
8. Dawkins, Richard. 1989. *The Selfish Gene.* Oxford: Oxford University Press.
9. Dennett, Daniel C. 2006. "Common-Sense Religion." *The Chronicle of Higher Education* 52(20):B6.
10. Harris, Sam. 2006. "Science Must Destroy Religion." *Edge,* January 2 (https://www.edge.org/response-detail/11122).

11. This quote—which has become a meme online—comes from an email exchange organized by *Newsweek* religion columnist Marc Gelman, in which Harris and syndicated talk show host Dennis Prager debated the question "Why are atheists so angry?" The exchange took place over four days and Harris's quote comes from an email written on November 16, 2006, with the subject heading "Yahweh Belongs on the Scrapheap of Mythology." See https://jewcy.com/jewish-religion-and-beliefs/monday_why_are_atheists_so_angry_sam_harris.

12. Numbers, Ronald L., and Jeff Hardin. 2018. "The New Atheists." In *The Warfare Between Science and Religion: The Idea That Wouldn't Die*. Edited by Jeff Hardin, Ronald L. Numbers, and Ronald A. Blinzley. Baltimore, MD: Johns Hopkins University Press, 220.

13. Keller, Tim. 2009. "Tim Keller on the New Atheists." *Big Think*. Accessed June 25, 2020, (https://bigthink.com/videos/tim-keller-on-the-new-atheists).

14. Schulson, Michael. 2014. "Jonathan Sacks on Richard Dawkins: 'New Atheists Lack a Sense of Humor.'" *Salon*. Accessed June 25, 2020, (https://www.salon.com/2014/09/27/jonathan_sacks_on_richard_dawkins_new_atheists_lack_a_sense_of_humor/).

15. See page 32 in Wood, Jack, and Gianpiero Petriglieri. 2005. "Transcending Polarization: Beyond Binary Thinking." *Transactional Analysis Journal* 35(1): 31–39.

16. Our survey had 679 atheists from the U.S. and 614 from the U.K.

17. Smith, Jesse M. 2013. "Creating a Godless Community: The Collective Identity Work of Contemporary American Atheists." *Journal for the Scientific Study of Religion* 52(1):80–81.

18. Habermas, Jurgen. 1971. *Knowledge and Human Interests*. Cambridge: Polity Press.

19. Mid-High/High SES Evangelical Church Houston Int5, conducted July 5, 2011.

20. See Ecklund, Elaine Howard, and Christopher P. Scheitle. 2017. *Religion vs Science: What Religious People Really Think*. New York: Oxford University Press; Johnson, David R., and Jared L. Peifer. 2017. "How Public Confidence in Higher Education Varies by Social Context." *The Journal of Higher Education* 88(4):619–644; Scheitle, Christopher P., and Elaine Howard Ecklund. 2017. "Recommending a Child Enter a STEM Career: The Role of Religion." *Journal of Career Development* 44(3):251–265.

21. Cornwall Alliance. 2010. "Sounding the Alarm about Dangerous Environmental Extremism: Explosive New DVD Series, Resisting The

Green Dragon, Now Being Distributed Nationally and Abroad." *Cornwall Alliance*. Accessed June 23, 2020, (https://cornwallalliance.org/2010/11/sounding-the-alarm-about-dangerous-environmental-extremism-explosive-new-dvd-series-resisting-the-green-dragon-now-being-distributed-nationally-and-abroad/).

22. Rodney Howard-Browne, the pastor of The River at Tampa Bay Church in Florida, for example, encouraged members of his church to shake hands with several people and said that only "pansies" would avoid such contact during the coronavirus, while Louisiana pastor Rev. Tony Spell continued to hold church services for more than 1,000 people in late March of 2020 and cast his church as a "hospital" where "anointed handkerchiefs" could heal individuals with COVID-19. See https://www.lgbtqnation.com/2020/03/pastor-laid-hands-trump-says-avoiding-coronavirus-pansies/ and https://www.newsweek.com/pastor-holds-service-over-1000-parishoners-defiance-large-gathering-ban-1493113.

23. Goldberg, Jonah. 2020. "The Treason of Epidemiologists." *The Dispatch*. Accessed June 23, 2020, (https://gfile.thedispatch.com/p/the-treason-of-epidemiologists).

24. Survey responses (RASIC U.K. Survey 2014) indicated that a significantly greater share of immigrants (48 percent) than non-immigrants (28 percent) reported to be affiliated with a religious tradition. More immigrants (25 percent) than non-immigrants (16 percent) also pray at least once a month, and a significantly smaller share of immigrants (29 percent) than non-immigrants (39 percent) believe that the relationship between science and religion is one of conflict.

25. Park, Alison, Caroline Bryson, Elizabeth Clery, John Curtice, and Miranda Philips. 2013. *British Social Attitudes: The 30th Report*. London: NatCen Social Research.

26. Rienzo, Cinzia, and Carlos Vargas-Silva. 2015. "Targeting Migration with Limited Control: The Case of the UK and the EU." *IZA Journal of European Labor Studies* 4(16).

27. While the growth of "nones" is an important aspect of the changing landscape of nonreligion, not all U.S. individuals in these groups are atheists. One survey of atheism in the U.K. found that 41.5 percent of religious nones believe that "there is definitely not a God or some 'higher power.'" See https://www.pewresearch.org/fact-tank/2015/05/13/a-closer-look-at-americas-rapidly-growing-religious-nones/.

28. As Zuckerman, Galen, and Pasquale (2016) explain, measuring what proportion of a population does not believe in God is complicated by a

number of challenges, including stigma associated with nonreligion, difficulties sampling small populations of nonreligious individuals, social desirability, and linguistic variation in how individuals construct secularity.

29. Pew Research Center, May 12, 2015. "America's Changing Religious Landscape." Sociologists Joseph Baker and Buster Smith analyzed data from the 2010 General Social Survey and also found that 3 percent of Americans are atheist. See Baker, Joseph O. and Buster G. Smith. 2015. *American Secularism: Cultural Contours of Nonreligious Belief Systems.* New York: Oxford University Press. To be sure, as one expands conceptualization of secularity beyond individuals who identify as atheists to include nonaffiliated believers and nonpracticing culturally religious individuals, the rough proportion of the American population who is secular is 28 percent.

30. Gervais, Will M., and Maxine Najile. 2018. "How Many Atheists Are There?" *Social Psychological and Personality Science* 9(1):3–10.

31. Lee, Lois. 2015. *Recognizing the Non-Religious.* Oxford: Oxford University Press.

32. Cragun, Ryan T., Barry Kosmin, Ariela Keysar, Joseph H. Hammer, and Michael Nielsen. 2012. "On the Receiving End: Discrimination Toward the Non-religious in the United States." *Journal of Contemporary Religion* 27(1):105–127. See also: Hammer, Joseph H., Ryan T. Cragun, Karen Hwang, and Jesse M. Smith. 2012. "Forms, Frequency, and Correlates of Perceived Anti-Atheist Discrimination." *Secularism and Nonreligion* 1:43–67; Doane, Michael J., and Marta Elliot. 2015. "Perceptions of Discrimination Among Atheists: Consequences for Atheist Identification, Psychological and Physical Well-Being." *Psychology of Religion and Spirituality* 7(2):130–142.

33. Encyclopedia Britannica. 2018. "United Kingdom." Accessed April 27, 2018, (http://www.britannica.com/EBchecked/topic/615557/United-Kingdom/44685/Religion); Sedghi, Ami. 2013. "UK Census: Religion by Age, Ethnicity and Country of Birth." *The Guardian*, May 16. And in the other U.K. nations outside of England—Scotland, Wales, and Northern Ireland—the Anglican Communion exerts significant social influence.

34. See Davie, Grace. 1990. "Believing without Belonging: Is This the Future of Religion in Britain?" *Social Compass* 37(4):455–469.

35. Jeffries, Stuart. 2007. "Britain's New Cultural Divide Is Not Between Christian and Muslim, Hindu, and Jew. It Is Between Those Who Have Faith and Those Who Don't." *The Guardian*, February 26. https://www.theguardian.com/world/2007/feb/26/religion.uk.

Chapter 2

1. Chesterton, G.K. 1910. "What's Wrong with the World." Mineola: Dover Publications. Pg. 29. The sub title "tried and found wanting" comes from this reference.
2. US27, female, graduate student, physics, conducted April 1, 2015.
3. Glass, Jennifer L., April Sutton, and Scott T. Fitzgerald. 2015. "Leaving the Faith: How Religious Switching Changes Pathways to Adulthood among Conservative Protestant Youth." *Social Currents* 2(2):126–143. See also Putnam, Robert D., and David E. Campbell. 2010. *American Grace: How Religion Divides and Unites Us*. New York: Simon & Schuster.
4. Mayrl, Damon, and Freeden Oeur. 2009. "Religion and Higher Education: Current Knowledge and Directions for Future Research." *Journal for the Scientific Study of Religion* 48(2):260–275.
5. Leuba, James H. 1934. "Religious Beliefs of American Scientists." *Harper's Magazine* 169:291–300.
6. Stark, Rodney. 1963. "On the Incompatibility of Religion and Science: A Survey of American Graduate Students." *Journal for the Scientific Study of Religion* 3:3–20.
7. Scholars also argue that part of this perceived incompatibility between science and religion may be due to social networks in elite research institutions rather than just knowledge about science itself. Elite institutions, and the social networks they facilitate, may indeed be as relevant or more relevant than learning more about science itself in shaping the religious attitudes and commitments of scientists. For example, sociologist Randall Collins argues that the academy plays a central role in shaping ideas about the development of worldviews. It is within university settings that academics form the kind of intimate social networks that help them become leaders in the transformation of culture. Elite university scientists then also have an important role in knowledge creation and institutional change because they provide scientific training to future societal leaders, but these same connections may make scientists and the students who study with them susceptible to cultural pressures to be irreligious as part of being a scientist. See Collins, Randall. 1998. *The Sociology of Philosophies: A Global Theory of Intellectual Change*. Cambridge, MA: Belknap Press of Harvard University Press.
8. Larson, Edward J., and Larry Witham. 1998. "Leading Scientists Still Reject God." *Nature* 394(313).

9. Stark, Rodney. 1963. "On the Incompatibility of Religion and Science: A Survey of American Graduate Students." *Journal for the Scientific Study of Religion* 3:3–20; Stark, Rodney, and Roger Finke. 2000. *Acts of Faith: Explaining the Human Side of Religion.* Berkeley and Los Angeles: University of California Press.

10. Stirrat, Michael, and R. Elizabeth Cornwell. 2013. "Eminent Scientists Reject the Supernatural: A Survey of the Fellows of the Royal Society." *Evolution: Education and Outreach* 6(1):33.

11. Ecklund, Elaine Howard, David R. Johnson, Christopher P. Scheitle, Kirstin R.W. Matthews, and Steven W. Lewis. 2016. "Religion among Scientists in International Context: A New Study of Scientists in Eight Regions." *Socius: Sociological Research for a Dynamic World* 2:1–9.

12. Gould, Stephen Jay. 1997. "Nonoverlapping Magisteria." *Natural History* 106:16–22.

13. Smith, Jesse M. 2011. "Becoming an Atheist in America: Constructing Identity and Meaning from the Rejection of Theism." *Sociology of Religion* 72(2):215–237, pg. 234.

14. UK108, male, postdoctoral fellow, biology, conducted March 1, 2011.

15. UK24, male, postdoctoral fellow, physics, conducted December 4, 2013.

16. UK130, female, graduate student, physics, conducted July 11, 2012.

17. UK130, female, graduate student, physics, conducted July 11, 2012.

18. UK35, female, senior lecturer, biology, conducted December 5, 2013.

19. UK108, male, postdoctoral fellow, biology, conducted March 1, 2011.

20. US36, female, graduate student, biology, conducted April 2, 2015.

21. US28, female, graduate student, physics, conducted April 1, 2015.

22. US68, female, professor, biology, conducted April 22, 2015.

23. UK03, female, principle investigator, biology, conducted December 2, 2013.

24. UK37, female, professor, biology, conducted December 5, 2013.

25. US36, female, Graduate Student, biology, conducted April 2, 2015.

26. US07, female, assistant professor, physics, conducted March 24, 2015.

27. UK79, female, senior research analyst, biology, conducted July 8, 2014.

28. UK109, male, reader, biology, conducted March 2, 2011.

29. UK45, female, professor, biology, conducted December 6, 2013.

30. UK54, female, graduate student, physics, conducted March 6, 2014.

31. UK01, male, graduate student, physics, conducted November 26, 2013.

32. UK02, male, graduate student, biology, conducted November 26, 2013.

33. UK22, male, lecturer, biology, conducted December 4, 2013.

34. UK26, female, lecturer, biology, conducted December 4, 2013.

35. Kosmin, Barry Alexander, and Ariela Keysar. 2009. *American Religious Identification Survey (ARIS 2008): Summary Report.* Trinity College.
36. In both countries, a majority of Asian atheists are nonnative. In the U.S. and U.K, 15 and 12 percent of Asian atheists were born in the country where they work, respectively.
37. See Appendix for an explanation of how we distinguish between elite and non-elite organizations in our sample.

Chapter 3

1. RASIC_UK113, biology, male, emeritus senior research fellow, conducted March 3, 2011.
2. RASIC_UK46, biology, female, professor, conducted December 6, 2013.
3. RASIC_UK24, physics, male, postdoc, conducted December 4, 2013.
4. Peterson, Gregory R. 2003. "Demarcation and the Scientistic Fallacy." *Zygon Journal of Religion and Science* 38(4):751–761.
5. RASIC_US01, biology, male, graduate student, conducted March 2, 2015.
6. RASIC_US97, biology, male, graduate student, conducted July 10, 2012.
7. RASIC_UK22, biology, male, lecturer, conducted December 4, 2013.
8. RASIC_UK108, biology, male, postdoctoral fellow, conducted March 1, 2011.
9. RASIC_US03, biology, female, graduate student, conducted March 2, 2015.
10. RASIC_US24, biology, male, professor, conducted April 1, 2015.
11. See Ecklund, Elaine Howard, David R. Johnson, Brandon Vaidyanathan, Kirstin R.W. Matthews, Steven W. Lewis, Robert A. Thomson Jr., and Di Di. 2019. *Secularity and Science: What Scientists around the World Really Think about Religion.* New York: Oxford University Press.
12. RASIC_US30, physics, female, associate professor, conducted April 2, 2015.
13. RASIC_UK39, biology, female, professor, conducted December 6, 2013.
14. RASIC_UK98, physics, female, senior research fellow, conducted October 27, 2014.
15. RASIC_UK42, physics, male, professor, conducted December 6, 2013.
16. RASIC_UK13, biology, female, reader, conducted December 3, 2013.
17. RASIC_US28, physics, female, graduate student, conducted April 1, 2015.
18. RASIC_US41, biology, female, professor, conducted April 3, 2015.
19. RASIC_UK41, biology, male, lecturer, conducted December 6, 2013.
20. 2010. "Ipsos Global @dvisory: Is Religion A Force For Good In The World? Combined Population of 23 Major Nations Evenly Divided in

Advance of Blair, Hitchens Debate." *Ipsos,* November 25. Accessed July 15, 2020 (https://www.ipsos.com/en-us/news-polls/ipsos-global-dvisory-religion-force-good-world-combined-population-23-major-nations-evenly-divided).

21. Pew Research Center. 2019. "Americans Have Positive Views About Religion's Role in Society, but Want It Out of Politics." *Pew Research Center,* November 15.

22. Harris, Sam. 2008. Pp. xii in *Letter to a Christian Nation.* New York: Vintage Books.

23. Hitchens, Christopher, Richard Dawkins, Sam Harris, and Daniel Dennett. 2019. *The Four Horsemen: The Conversation that Sparked an Atheist Revolution.* New York: Random House.

24. Weinberg, Steven. 1999. "A Designer Universe?" *Conference on Cosmic Design of the American Association for the Advancement of Science.* Accessed June 29, 2020 (https://www.physlink.com/Education/essay_weinberg.cfm).

25. Stenger, Victor. 2007. *God: The Failed Hypothesis—How Science Shows that God Does Not Exist.* Amherst, MA: Prometheus Books.

26. RASIC_US35, physics, female, graduate student, conducted April 2, 2015.

27. RASIC_US35, physics, female, graduate student, conducted April 2, 2015.

28. Parsons, Talcott, and Gerald Platt. 1968. In *The American Academic Profession: A Pilot Study.* National Science Foundation, pp. 1–14.

29. Du Bois, W.E.B. 1940. *Dusk of Dawn: An Essay Toward An Autobiography of a Race Concept,* 130–133. New York: Harcourt Brace.

30. RASIC_UK22, biology, male, lecturer, conducted December 4, 2013.

31. RASIC_US24, biology, male, professor, conducted April 1, 2015.

32. RASIC_UK114, biology, male, professor, conducted March 4, 2011.

33. RASIC_UK10, physics, female, postdoctoral fellow, conducted December 2, 2013.

34. Cf. Haberman, Clyde. 2014. "From Private Ordeal to National Fight: The Case of Terri Schiavo." *New York Times,* April 20. Accessed July 1, 2020 (https://www.nytimes.com/2014/04/21/us/from-private-ordeal-to-national-fight-the-case-of-terri-schiavo).

35. RASIC_UK80, biology, female, senior lecturer, conducted July 8, 2014.

36. RASIC_UK102, biology, male, research professor, conducted February 28, 2011.

37. RASIC_US88, biology, female, postdoc, conducted March 19, 2012.

38. RASIC_UK54, physics, female, graduate student, conducted March 6, 2014.

39. There is a sizeable body of research that examines the relationship between religion and charitable giving. See for example: Vaidyanathan, Brandon, Jonathan P. Hill, and Christian Smith. 2011. "Religion and Charitable Financial Giving to Religious and Secular Causes: Does Political Ideology Matter?" *Journal for the Scientific Study of Religion* 50(3):450–469; Hill, Jonathan P., and Brandon Vaidyanathan. 2011. "Substitution or Symbiosis? Assessing the Relationship Between Religious and Secular Giving." *Social Forces* 90(1):157–180; and Schnable, Allison. 2015. "Religion and Giving for International aid: Evidence From a Survey of US Church Members." *Sociology of Religion* 76(1): 72–94.

40. RASIC_UK117, biology, male, professor, conducted March 8, 2011.

41. RASIC_UK98, physics, female, senior research fellow, conducted October 27, 2014.

42. RASIC_UK46, biology, female, professor, conducted December 6, 2013.

43. RASIC_US19, physics, graduate student, conducted March 27, 2015.

44. RASIC_US89, biology, male, graduate student, conducted June 28, 2012.

45. Ecklund et al. 2019. [*Secularity and Science*. Already cited in book]

46. Dawkins, Richard. 2006. Pp. 204–206 in *The God Delusion*. London: Bantam Press.

47. Simon, Scott. 2017. "Richard Dawkins on Terrorism and Religion." *NPR*, May 27.

48. Dawkins, Richard. 2006. Pp. 259 in *The God Delusion*. London: Bantam Press.

49. Dawkins, Richard. 2006. Pp. 31 in *The God Delusion*. London: Bantam Press.

50. Dawkins, Richard. 2006. Pp. 126 in *The God Delusion*. London: Bantam Press.

51. RASIC_UK03, biology, female, principal investigator and group leader, conducted December 2, 2013.

52. RASIC_UK07, biology, male, professor, conducted December 2, 2013.

53. RASIC_UK13, biology, female, reader, conducted December 3, 2013.

54. RASIC_UK02, biology, male, graduate student, conducted November 26, 2013.

55. Johnson, David, Elaine Howard Ecklund, Di Di, and Kirstin Matthews. 2018. "Responding to Richard: Celebrity and (Mis)representation of Science." *Public Understanding of Science* 27(5):535–549.

56. RASIC_UK08, biology, male, professor, conducted December 2, 2013.

57. RASIC_UK114, biology, male, professor, conducted March 4, 2011.

58. RASIC_UK108, biology, male, postdoctoral fellow, conducted March 1, 2011.

59. RASIC_UK41, biology, male, lecturer, conducted December 6, 2013.

Chapter 4

1. See https://www.focusonthefamily.com/family-qa/marriage-between-an-atheist-and-a-christian/. Italics in the original.

2. A considerable amount of research has pointed to Reddit as an important venue for atheists. The website also has been the basis of sampling atheists for some researchers. Zuckerman et al. (p. 220) describe it as the new "atheist agora" and discuss its role as an important public sphere in which nonreligious individuals of all types communicate, connect, and form groups. Reddit is just one of many internet venues for atheists, alongside Facebook, Meetup, and the websites of atheist organizations.

3. https://www.reddit.com/r/atheism/comments/56munj/atheist_married_to_christian/.

4. Smith, Jessie M. 2011. "Becoming an Atheist in America: Constructing Identity and Meaning from the Rejection of Theism." *Sociology of Religion* 72(2):21–237, pg. 228.

5. For more on homophily, see: Marsden, Peter V. 1988. "Homogeneity in Confiding Relations." *Social Networks* 10:57–76; and McPherson, J. Miller, and Lynn Smith-Lovin. 1987. "Homophily in Voluntary Organizations: Status Distance and the Composition of Face to Face Groups." *American Sociological Review* 52:370–379. The general level of segregation one observes between religious individuals and modernist and spiritual atheists is more likely driven by opportunity structure.

6. McPherson, J. Miller, Lynn Smith-Lovin, and James M. Cook. 2001. "Birds of a Feather: Homophily in Social Networks." *Annual Review of Sociology* 27:415–444.

7. RASIC_UK34, biology, male, professor, conducted December 5, 2013.

8. Our usage of the term *culturally religious* draws on existing research on atheism. Garelli (2016a: vii) emphasizes that atheists in this category "have been socialized within a religious environment but consider religion as a merely cultural phenomenon with no commitment to belief." We expand this conceptual framing, however, to include atheists who have sustained interactions with or commitments to religious individuals or religious organizations. This does not make them religious in a conventional sense, but it calls attention to an active decision to expose one's

self or immediate family members to religious culture. For other research on cultural atheists, see: Beaman, Lori, and Steven Tomlins (Eds). 2015. *Atheist Identities: Spaces and Social Contexts.* Volume 2 in *Boundaries of Religious Freedom: Regulating Religion in Diverse Societies* book series. Switzerland: Springer; Demarath III, N. J. 2000. "The Rise of 'Cultural Religion' in European Christianity: Learning from Poland, Northern Ireland, and Sweden." *Social Compass* 47(1):127–139; Garelli, Franco. 2014. *Religion Italian Style: Continuities and Change in a Catholic Community.* Abingdon-on-Thames, UK: Routledge; Garelli, Franco. 2016. "Preface." In *Annual Review of the Sociology of Religion: Sociology of Atheism,* edited by R. Cipriani and F. Garelli. Leiden, Netherlands: Brill; Zuckerman, Phil. 2012. *Faith No More: Why People Reject Religion.* New York: Oxford University Press.

9. RASIC_UK99, biology, male, senior lecturer, conducted February 28, 2011.
10. The notion of "believing without belonging" was coined by sociologist Grace Davie. The phrase has taken on a life of its own beyond her claim. Some proponents of the idea assert that belief in the sacred is high while religious practice has declined. Other formulations depict "believing without belonging" as either involving alternative spiritualities (widening the definition of religion) that do not entail formal practices, or as a societal transition toward secularism. Sociologists David Voas and Alasdair Crockett, drawing on data from the British Household Panel Survey and the British Social Attitudes Survey, provide evidence that supports this latter emphasis: belief has eroded in Britain at the same rate as religious affiliation and attendance. See Davie, Grace. 1990. "Believing Without Belonging: Is This the Future of Religion in Britain?" *Social Compass* 37(4):455–469; and Voas, David, and Alasdair Crockett. 2005. "Religion in Britain: Neither Believing nor Belonging." *Sociology* 39(1):11–28.
11. Among current atheists, 11 percent had an affiliation at age 16 and 3 percent had no affiliation at that age but claim one now.
12. The relative absence of Muslims who identify as such but without belief may reflect the deeply complicated nature of leaving Islam more generally. In Simon Cottee's (2015) *The Apostates: When Muslims Leave Islam,* he shows that anyone who would seek to leave Islam faces major constraints. Based on interviews with 35 ex-Muslims from the UK and Canada, he finds that individuals who leave Islam often encounter rejection, ostracism, and criticism from family, friends, and leaders within local Islamic communities. Such challenges only compound identity and other personal changes that accompany leaving a religious community, whether a "strict

religion" or otherwise. See Cottee, Simon. 2015. *The Apostates: When Muslims Leave Islam*. London: Hurst & Company.

13. RASIC UK 17, theoretical astrophysics, male, professor, conducted December 3, 2013.

14. RASIC UK137, physics, female, professor, conducted July 13, 2012.

15. Cohen, Steven M., and Arnold M. Eisen. 2000. *The Jew Within: Self, Family, and Community in America*. Bloomington: Indiana University Press, pg. 91. This choice is only emboldened by a shared sense of history that binds people to a religious community. Indeed, some sociological research finds that many contemporary Jews embrace an essentialist view of Jewishness that emphasizes the biological/genetic basis of identity (Tenenbaum and Davidman, 2007)—even as scientists and scholars have abandoned such racial categorizations (Goldstein 1997). Such work underscores the notion of a nontheological basis of Jewish identity. Tenenbaum, Shelly, and Lynn Davidman. 2007. "It's in My Genes: Biological Discourse and Essentialist Views of Identity Among Contemporary American Jews." *The Sociological Quarterly* 48(3):435–450; and Goldstein, Eric L. 1997. "'Different Blood Flows in Our Veins': Race and Jewish Self-Definition in Late-Nineteenth-Century America." *American Jewish History* 85:29–55.

16. As we note, becoming an atheist within a religious community is easier in some traditions relative to others. As Cottee (2015) observes, the process for Muslims is especially complicated. While we did not interview any atheist Muslim scientists, existing work does point to different ways that some Muslims mix secularity with religious tradition. Martin (2010) argues that there are three ways some Muslims maintain their Muslim identity while embracing secularity. One category entails "hard" Islamic secularists who engage intellectually and socially with Islam and who feel a sense of ownership of the cultural achievements of Islamic civilization. Another category is the "soft" Muslim secularist, who is silent about his or her lack of religious commitment, but maintains a political identity tied to Islam, such as opposing theocratic governments without openly denying religion in the way that a hard secularist would. Finally, Martin emphasizes a third category of political secularists who are not personally religious, but believe that a public policy of secularism is important for all religious communities. See Martin, Richard C. 2010. "Hidden Bodies in Islam: Secular Muslim Identities in Modern (and Premodern) Societies." In *Muslim Societies and the Challenge of Secularization: An Interdisciplinary Approach*, edited by G. Marranci. *Muslims in Global Societies Series* 1, pp. 131–148, DOI 10.1007/978-90-481-3362-8_9.

17. Dein, Simon. 2013. "The Origins of Jewish Guilt: Psychological, Theological, and Cultural Perspectives." *Journal of Spirituality in Mental Health* 15(2):123–137.

18. See, for example, Rubenstein, Richard. 1966. *After Auschwitz: Radical Theology and Contemporary Judaism*. Indianapolis: Bobbs-Merrill.

19. RASIC_UK135, physics, male, professor, conducted July 13, 2012.

20. RASIC_US26, physics, female, graduate student, conducted April 1, 2015.

21. RASIC UK135, physics, male, professor, conducted July 13, 2012.

22. RASIC_ US36, biology, female, graduate student, conducted April 2, 2015.

23. RASIC_US53, physics, male, professor, conducted April 2, 2015.

24. RASIC_US53, physics, male, professor, conducted April 2, 2015.

25. RASIC_ UK100, biology, female, senior research associate, conducted February 28, 2011.

26. RASIC_UK99, biology, male, senior lecturer, conducted February 28, 2011.

27. Collins, Randall. 1993. "Emotional Energy as the Common Denominator of Rational Action." *Rationality and Society* 5(2):203–230, pg. 205.

28. Bourdieu, Pierre. 1986. "The Forms of Capital." In *Handbook of Theory and Research for the Sociology of Education*, edited by J. Richardson. New York: Greenwood, pp. 46–58.

29. Bourdieu, Pierre. 1973. "Cultural Reproduction and Social Reproduction." In *Knowledge, Education and Cultural Changes*, edited by R. Brown. London: Tavistock, pp. 71–112.

30. RASIC_UK34, biology, male, professor, conducted December 5, 2013.

31. Bourdieu, Pierre. 1977. "Cultural Reproduction and Social Reproduction." In *Power and Ideology in Education*, edited by J. Karabel and A.H. Halsey. New York: Oxford University Press.

32. Lareau, Annette. 2003. *Unequal Childhoods: Class, Race, and Family Life*. Berkeley: University of California Press.

33. Hemming, Peter J., and Christopher Roberts. 2018. "Church Schools, Educational Markets and the Rural Idyll." *British Journal of Sociology of Education* 39(4):501–517.

34. RASIC_UK34, biology, male, professor, conducted December 5, 2013.

35. RASIC_UK34, biology, male, professor, conducted December 5, 2013.

36. RASIC_UKO7, biology, male, professor, conducted December 2, 2013.

37. RASIC_UK17, theoretical astrophysics, male, professor, conducted December 3, 2013.

38. Davie, Grace. 2007. *The Sociology of Religion*. London: Sage.

39. RASIC_UK14, physics, male, professor, conducted December 3, 2013.

40. RASIC_UK31, physics, male, professor, conducted December 5, 2013.
41. RASIC_US06, physics, female, graduate student, conducted March 24, 2015.
42. RASIC_US02, biology, male, graduate student, conducted March 2, 2015.
43. Dawkins, Richard. 2015. "Don't Force Your Religious Opinions on Your Children." *Time*, February 19.
44. Di Fiore, James. 2016. "My Kids Will Be Raised Without Religion." *Huffpost*, April 25.
45. RASIC_UK14, physics, male, professor, conducted December 3, 2013.
46. RASIC UK17, theoretical astrophysics, male, professor, conducted December 3, 2013.

Chapter 5

1. RASIC_UK41, biology, male, lecturer, conducted December 6, 2013.
2. Berger, Peter. 2014. *The Many Altars of Modernity: Toward a Paradigm for Religion in a Pluralist Age*. Berlin: Walter de Gruyter GmbH & Co KG.
3. Lipka, Michael, and Claire Gecewicz. 2017. "More Americans Now Say They're Spiritual but Not Religious." *Pew Research Center*, September 6. Accessed July 13, 2020 (https://www.pewresearch.org/fact-tank/2017/09/06/more-americans-now-say-theyre-spiritual-but-not-religious/); Pew Research Center. 2018. "Attitudes Toward Spirituality and Religion" in "Being Christian in Western Europe." *Pew Research Center*, May 29. Accessed July 15, 2020 (https://www.pewforum.org/2018/05/29/attitudes-toward-spirituality-and-religion/).
4. Bellah, Robert, Richard Madsen, William M. Sullivan, Ann Swidler, and Steven M. Tipton. 1985. *Habits of the Heart: Individualism and Commitment in American Life*. New York: Harper & Row; Dillon, Michele, and Paul Wink. 2007. *In the Course of a Lifetime: Tracing Religious Belief, Practice, and Change*. Berkeley: University of California Press; Heelas, Paul, and Linda Woodhead. 2005. *The Spiritual Revolution: Why Religion Is Giving Way to Spirituality*, edited by B. Seel, B. Szerszynski, and K. Tusting. Malden, MA: Blackwell Publishing; Wuthnow, Robert. 1998. *After Heaven: Spirituality in America since the 1950s*. Berkeley and Los Angeles: University of California Press.
5. Marler, Penny Long, and C. Kirk Hadaway. 2002. "'Being Religious' or 'Being Spiritual' in America: A Zero-Sum Proposition?" *Journal for the Scientific Study of Religion* 41:289–300; Underwood, Lynn G., and Jeanne A. Teresi. 2002. "The Daily Spiritual Experience Scale: Development, Theoretical

Description, Reliability, Exploratory Factor Analysis, and Preliminary Validity Using Health-Related Data." *Annual Behavioral Medicine* 24:22–33; Watson, P. J., and Ronald Morris. 2005. "Spiritual Experience and Identity: Relationships with Religious Orientation, Religious Interest, and Intolerance of Ambiguity." *Review of Religious Research* 46:371–379.

6. Roof, Wade Clark. 1998. "Religious Borderlands: Challenges for Future Study." *Journal for the Scientific Study of Religion* 37:1–14; Roof, Wade Clark. 1999. *Spiritual Marketplace: Baby Boomers and the Remaking of American Religion.* Princeton, NJ: Princeton University Press; Wuthnow, Robert, and Wendy Cadge. 2004. "Buddhists and Buddhism in the United States: The Scope of Influence." *Journal for the Scientific Study of Religion* 43:363–380.

7. Sahgal, Neha. 2018. "10 Key Findings about Religion in Western Europe." *Pew Research Center*, May 29. Accessed July 15, 2020 (https://www.pewresearch.org/fact-tank/2018/05/29/10-key-findings-about-religion-in-western-europe/).

8. Berger, Peter. 2014. *The Many Altars of Modernity: Toward a Paradigm for Religion in a Pluralist Age.* Berlin: Walter de Gruyter GmbH & Co KG.

9. Bruce, Steve. 2002. *God Is Dead: Secularization in the West.* Malden, MA: Wiley-Blackwell; Heelas, Paul. 2006. "Challenging Secularization Theory: The Growth of 'New Age' Spiritualties of Life." *Hedgehog Review* 8(1):46.

10. Aupers, Stef and Dick Houtman. 2006. "Beyond the Spiritual Supermarket: The Social and Public Significance of New Age Spirituality." *Journal of Contemporary Religion* 21(2):201–222.

11. Draper, Scott, and Joseph O. Baker. 2011. "Angelic Belief as American Folk Religion." *Sociological Forum* 26(3):623–643; Eaton, Marc A. 2015. "'Give Us a Sign of Your Presence': Paranormal Investigation as a Spiritual Practice." *Sociology of Religion* 76(4):389–412.

12. See Mercandante, Linda A. 2014. *Belief Without Borders: Inside the Minds of the Spiritual but Not Religious.* New York: Oxford University Press, pg. 93.

13. Heelas, Paul, and Linda Woodhead 2005. *The Spiritual Revolution: Why Religion Is Giving Way to Spirituality*, edited by B. Seel, B. Szerszynski, and K. Tusting. Malden, MA: Blackwell Publishing; Kraus, Rachel. 2009. "Straddling the Sacred and Secular: Creating a Spiritual Experience Through Belly Dance." *Sociological Spectrum* 29(5):598–625.

14. Draper, Scott, and Joseph O. Baker. 2011. "Angelic Belief as American Folk Religion." *Sociological Forum* 26(3):623–643; Eaton, Marc A. 2015.

"'Give Us a Sign of Your Presence': Paranormal Investigation as a Spiritual Practice." *Sociology of Religion* 76(4):389–412.

15. Ecklund, Elaine Howard, and Elizabeth Long. 2011. "Scientists and Spirituality." *Sociology of Religion* 72(3):253–274.

16. Ammerman, Nancy T. 2013. "Spiritual But Not Religious? Beyond Binary Choices in the Study of Religion." *Journal for the Scientific Study of Religion* 52(2):258–278.

17. Ecklund, Elaine Howard, and Elizabeth Long. 2011. "Scientists and Spirituality." *Sociology of Religion* 72(3):253–274.

18. Wuthnow, Robert. 2002. *Loose Connections*. Cambridge: Harvard University Press, pp. 22–23.

19. Our data do not permit a concrete explanation for why these differences emerge. Part of the difference is driven by the prevalence of the independence view in each context. We find that 81 percent of the spiritual atheists in the U.K. and 56 percent of spiritual atheists in the U.S. see science and religion as different aspects of reality. One speculation is the greater prevalence of conflict in the public sphere in the U.S. contributes to more personal adherence to the conflict view among spiritual atheists.

20. See Ecklund, Elaine Howard. 2010. *Science Vs. Religion: What Scientists Really Think*. New York: Oxford University Press.

21. RAAS_Bio9, conducted July 25, 2005.

22. RASIC_US30, biology, female, associate professor, conducted April 2, 2015.

23. Ecklund, Elaine Howard, and Elizabeth Long. 2011. "Scientists and Spirituality." *Sociology of Religion* 72(3):253–274.

24. RASIC_UK41, biology, male, lecturer, conducted December 6, 2013.

25. RASIC_UK41, biology, male, lecturer, conducted December 6, 2013.

26. RASIC_UK20, physics, male, reader, conducted December 3, 2013.

27. RASIC_UK35, biology, female, senior lecturer, conducted December 5, 2013.

28. RASIC_UK20, physics, male, reader, conducted December 3, 2013.

29. RASIC_UK35, biology, female, senior lecturer, conducted December 5, 2013.

30. Ecklund, Elaine Howard, and Elizabeth Long. 2011. "Scientists and Spirituality." *Sociology of Religion* 72(3):253–274.

31. Bartkowski, John. 2004. *The Promise Keepers: Servants, Soldiers, and Godly Men*. New Brunswick, NJ: Rutgers University Press; Lamont, Michè`le. 2001. "Culture and Identity." In *Handbook of Sociological Theory*, edited by Jonathan H. Turner. New York: Plenum, pp. 171–186.

32. RAAS Phys5, conducted July 12, 2005.

33. RASIC_US41, biology, female, professor, conducted April 3, 2015.

34. RASIC_US97, biology, male, graduate student, conducted July 10, 2012.

35. Ecklund, Elaine Howard, and Elizabeth Long. 2011. "Scientists and Spirituality." *Sociology of Religion* 72:253–274.

36. RASIC_UK114, biology, male, professor, conducted March 4, 2011.

37. RASIC_US27, physics, female, graduate student, conducted April 1, 2015.

38. Hamburger, Philip. 2009. *Separation of Church and State.* Cambridge, MA: Harvard University Press.

39. Beaman, Lori, and Steven Tomlins. 2015. *Atheist Identities: Spaces and Social Contexts.* New York: Springer International Publishing; Egdell, Penny et al. 2006. "Atheists as 'Other': Moral Boundaries and Cultural Membership in American Society." *American Sociological Review* 72(2):211–234.

40. Beaman, Lori G., and Steven Tomlins. 2014. *Atheist Identities: Spaces and Social Contexts.* New York: Springer International Publishing.

41. RASIC_US60, biology, female, associate professor, conducted April 15, 2015.

42. RASIC_UK45, biology, female, professor, conducted December 6, 2013.

43. RASIC_UK09, biology, female, lecturer, conducted December 2, 2013.

44. RASIC_US60, biology, female, associate professor, conducted April 15, 2015.

45. See McMahan, David L. 2004. "Buddhism: Introducing the Buddhist Experience (review)." *Philosophy East and West* 54(2):268–270.

46. RASIC_UK114, biology, male, professor, conducted March 4, 2011.

47. RASIC_UK28, physics, male, postdoctoral fellow, conducted December 4, 2013.

48. RAAS Psyc15, conducted October 19, 2005.

49. RASIC_UK114, biology, male, professor, conducted March 4, 2011.

50. RASIC_US60, biology, female, associate professor, conducted April 15, 2015.

Chapter 6

1. Connor, Steve. 2011. "For the Love of God . . . Scientists in Uproar at £1m Religion Prize." *The Independent*, April 7. Accessed January 15, 2015 (http://www.independent.co.uk/news/science/for-the-love-of-god-scientists-in-uproar-at-1631m-religion-prize-2264181.html).

2. RASIC_UK07, biology, professor, male, conducted December 2, 2013.

3. RASIC_UK45, biology, professor, female, conducted December 6, 2013.

4. Our concern in this chapter, in part, is how a variety of atheist scientists make sense of the limits of science. For in-depth accounts of how individual religious scientists think about these limits, see: Collins, Francis. 2006. *The Language of God: A Scientist Presents Evidence for Belief*. New York: Free Press; and Gleiser, Marcelo. 2014. *The Island of Knowledge: The Limits of Science and the Search for Meaning*. New York: Basic Books.

5. Merton, Robert K. 1957. "Priorities in Scientific Discovery: A Chapter in the Sociology of Science." *American Sociological Review* 22(6):635–659.

6. One of the central mechanisms for upholding objectivity in science is peer review, which seeks to ensure that only scientific criteria govern what "counts" as socially certified knowledge. A wealth of research, however, has identified dysfunctions within the peer review process which undermine objectivity in the production of knowledge. Reviewers of scientific research, for example, may be biased by particularistic factors unrelated to the production of knowledge such as the institutional affiliation or gender of the researcher. For an intensive review of this literature, see Johnson, David R., and Joseph C. Hermanowicz. 2017. "Peer Review: From 'Sacred Ideals' to 'Profane Realities.'" In *Higher Education: Handbook of Theory and Research*. Switzerland: Springer, pp. 485–527.

7. Kierkegaard, Soren. 1846. *Concluding Postscripts to Scientific Inquiry* in *Philosophical Fragments*. Tr. Howard V. Hong and Edna H. Hong, 1992. Princeton, NJ: Princeton University Press.

8. Goffman, Erving. 1963. *Behavior in Public Places*. Glencoe, IL: Free Press of Glencoe.

9. Ecklund, Elaine Howard. 2010. *Science Vs. Religion: What Scientists Really Think*. New York: Oxford University Press.

10. Johnson, David R. 2017. In *A Fractured Profession: Commercialism and Conflict in Academic Science*. Baltimore, MD: Johns Hopkins University Press, pp. 47–48.

11. Consider the comment of Richard Horton, a former editor of the prestigious medical journal *The Lancet*. As he writes, "Editors and scientists alike insist on the pivotal importance of peer review. We portray peer review to the public as a quasi-sacred process that helps make science our most objective truth teller. But we know that the system of peer review is biased, unjust, unaccountable, incomplete, easily fixed, often insulting, usually ignorant, occasionally foolish, and frequently wrong." Horton, Richard. 2000. "Genetically Modified Food: Consternation, Confusion, and Crack-Up." *Medical Journal of Australia* 177:148–149.

12. Ben-David, Joseph, and Teresa A. Sullivan. 1975. In "Sociology of Science." *Annual Review of Sociology* 1(1): 203–222, pg. 203.

13. Gieryn, Thomas F. 1999. In *Cultural Boundaries of Science: Credibility on the Line*. Chicago: University of Chicago Press, pg. x.

14. Gieryn, Thomas F. 1999. *Cultural Boundaries of Science: Credibility on the Line*. Chicago: University of Chicago Press.

15. Gieryn, Thomas F. 1999. In *Cultural Boundaries of Science: Credibility on the Line*. Chicago: University of Chicago Press, pg. 10.

16. See, in particular, David's work on this topic: Johnson, David R. 2017. *A Fractured Profession: Commercialism and Conflict in Academic Science*. Baltimore, MD: Johns Hopkins University Press.

17. RASIC_UK02, biology, graduate student, male, conducted November 26, 2013.

18. RASIC_UK23, physics, lecturer, male, conducted December 4, 2014.

19. RASIC_UK40, biology, research fellow, male, conducted December 6, 2013.

20. Twelve atheist scientists we interviewed expressed this belief. Of these 12, seven were modernists, three were spiritual atheists, and two were culturally religious.

21. RASIC_US37, biology, graduate student, male, conducted April 2, 2015.

22. RASIC_UK09, biology, lecturer, female, conducted December 2, 2013.

23. Quoted in Merton, Robert K. 1973. In *The Sociology of Science: Theoretical and Empirical Investigations*. Chicago: University of Chicago Press, pg. 303.

24. Mitroff, Ian. 1974. "Norms and Counter-Norms in a Select Group of the Apollo Moon Scientists: A Case Study of the Ambivalence of Scientists." *American Sociological Review* 39(4):579–595.

25. RASIC_US03, biology, graduate student, female, conducted March 2, 2015.

26. RASIC_US01, biology, graduate student, male, conducted March 2, 2015.

27. See, for example, Ammerman, Nancy. 2013. *Sacred Stories, Spiritual Tribes: Finding Religion in Everyday Life*. New York: Oxford University Press. See also, Wuthnow, Robert. 2020. *What Happens When We Practice Religion? Textures of Devotion in Everyday Life*. Princeton, NJ: Princeton University Press.

28. RASIC_UK54, physics, graduate student, female, conducted March 6, 2014.

29. RASIC_US41, biology, professor, female, conducted April 3, 2015.

30. RASIC_UK41, biology, lecturer, male, conducted December 6, 2013.

31. See Ecklund, Elaine Howard, and Elizabeth Long. 2011. "Scientists and Spirituality." *Sociology of Religion* 72(3):253–274.

32. RASIC_UK20, physics, reader, male, conducted December 3, 2013.

33. RASIC_UK98, physics, senior research fellow, female, conducted October 27, 2014.

34. Dawkins, Richard. 2011. "The Virus of Faith." Documentary UK Channel 4. Accessed October 21, 2019 (https://www.youtube.com/watch?v=eVy-0E1x620).

35. See, for example, Ecklund, Elaine Howard. 2010. *Science vs. Religion: What Scientists Really Think*. New York: Oxford University Press.

36. As an ideology, scientism lacks a formal doctrine, even as philosophers have debated and delineated different strains of it (cf. Stenmark, Mikael. 2001. *Scientism: Science, Ethics, and Religion*. Aldershot, UK: Ashgate; De Ridder, Jeroen, Rik Peels, and Rene van Woudenberg (Eds.) 2018. *Scientism: Prospects and Problems*. Oxford, U.K.: Oxford University Press; and Boudry, Maarten, and Massimo Pigliucci (Eds.) 2017. *Science Unlimited? The Challenges of Scientism*. Chicago: University of Chicago Press. Here we borrow the phrase "scientism wars" from an essay by Oliver Burkeman in *The Guardian*. See Burkeman, Oliver. 2013. "'Scientism' Wars: There's an Elephant in the Room, and Its Name Is Sam Harris." August 27. *The Guardian*. Accessed July 24, 2020 (https://www.theguardian.com/news/oliver-burkeman-s-blog/2013/aug/27/scientism-wars-sam-harris-elephant).

37. Atkins, Peter. 1981. *The Creation*. New York: W.H. Freeman & Co, pg. 3.

38. Hawking, Stephen. 1996. *A Brief History of Time: The Updated and Expanded Tenth Anniversary Edition*. New York: Bantam Books, pg. 14.

39. In the book, *The Grand Design*, Hawking and his colleague Leonord Mlodinow wrote that "philosophy is dead. . . . Scientists have become the bearers of the torch of discovery in our quest for knowledge." (p. 5) In a podcast interview, Neil deGrasse Tyson depicted philosophy as useless when he stated that " . . . [T]he philosophers believe they are actually asking deep questions about nature. And to the scientist, it's what are you doing? Why are you concerning yourself with the meaning of meaning?" Bill Nye, in an interview for a web series called *Big Think*, stated that "philosophy is important for a while but . . . you start arguing in a circle where I think therefore I am. What if you don't think about it? Do you not exist anymore? You probably still exist even if you're not thinking about existence." See https://scientiasalon.wordpress.com/2014/05/12/neil-degrasse-tyson-and-the-value-of-philosophy/; https://www.youtube.com/watch?v=ROe28Ma_

tYM&feature=youtu.be; and Hawking, Stephen, and Leonard Mlodinow. 2010. *The Grand Design*. New York: Bantam Books.

40. RASIC_UK79, biology, senior research associate, female, conducted July 8, 2014.

41. RASIC_UK22, biology, lecturer, male, conducted December 4, 2013.

42. RASIC_UK24, physics, postdoc, male, conducted December 4, 2013.

43. Crick, Francis. 1994. *The Astonishing Hypothesis: The Scientific Search for the Soul*. New York: Touchstone.

44. Sagan, Carl. 1980. *Cosmos*. New York: Ballantine Books, pg. 105.

45. Rosenberg, Alex. 2011. *The Atheist's Guide to Reality: Enjoying Life Without Illusions*. New York: W.W. Norton & Company, pg. 8.

46. RASIC_US07, physics, assistant professor, female, conducted March 24, 2015.

47. RASIC_UK37, biology, professor, female, conducted December 5, 2013.

48. Crick, Francis. 1994. *The Astonishing Hypothesis: The Scientific Search for the Soul*. New York: Touchstone, pg. 3.

49. Sagan, Carl. 1980. *Cosmos*. New York: Ballantine Books, pg. 105.

50. RASIC_UK46, biology, professor, female, conducted December 6, 2013.

51. RASIC_UK26, biology, lecturer, female, conducted December 4, 2013.

52. RASIC_UK39, biology, professor, female, conducted December 6, 2013.

53. RASIC_UK98, physics, senior research fellow, female, conducted October 27, 2014.

54. RASIC_UK03, biology, principal investigator, female, conducted December 2, 2013.

55. RASIC_UK03, biology, principal investigator, female, conducted December 2, 2013.

56. RASIC_UK20, physics, reader, male, conducted December 3, 2013.

57. RASIC_UK36, biology, professor, male, conducted December 5, 2013.

58. RASIC_US02, biology, graduate student, male, conducted December 2, 2015.

59. Wilson, Edward O. 1978. *On Human Nature*. Cambridge, MA: Harvard University Press, pg. 5.

60. Harris, Sam. 2010. *The Moral Landscape: How Science Can Determine Human Values*. New York: Free Press, pg. 28.

61. RASIC_US19, physics, graduate student, male, conducted December 25, 2015.

62. RASIC_US26, physics, graduate student, female, conducted April 1, 2015.

63. RASIC_US02, biology, graduate student, male, conducted March 2, 2015.

Chapter 7

1. RASIC_UK121, physics, professor, male, conducted March 7, 2012.

2. Cf. Genesis 1:27, New American Standard Bible, and subsequent discussion in Millard, Erickson J. 1983. *Christian Theology*. Grand Rapids, MI: Baker Academic, Chapter 22. Also Wright, N.T. 2013. *Paul and the Faithfulness of God*. Minneapolis: Fortress Press. See 406, 441, 640 and footnotes for fuller discussion of original intent and meaning interpreted by early Christian authors.

3. Wuthnow, Robert. 1989. *Communities of Discourse: Ideology and Social Structure in the Reformation, the Enlightenment, and European Socialism*. Cambridge, MA: Harvard University Press.

4. For a detailed examination of secular alternatives to religion in answering questions of meaning, see: Blessing, Kimberly. 2013. "Atheism and the Meaningfulness of Life." In *The Oxford Handbook of Atheism*, edited by Stephen Bullivant and Michael Ruse. New York: Oxford University Press, pp. 104–118; Ruse, Michael. 2019. *A Meaning to Life*. New York: Oxford University Press.

5. Blessing, Kimberly. 2013. "Atheism and the Meaningfulness of Life." In *The Oxford Handbook of Atheism*, edited by Stephen Bullivant and Michael Ruse. New York: Oxford University Press, pp. 104–118; pg. 105.

6. Frost, Jacqui. 2019. "Certainty, Uncertainty, or Indifference? Examining Variation in the Identity Narratives of Nonreligious Americans." *American Sociological Review* 84(5):828–850.

7. Frost, Jacqui. 2017. "Rejecting Rejection Identities: Negotiating Positive Non-religiosity at the Sunday Assembly." In *Religion and Its Others: Studies in Religion, Nonreligion, and Secularity, Volume 6*, edited by Stacey Gutkowski, Lois Lee, and Johannes Quack. Boston and Berlin: Walter de Gruyter.

8. For a detailed discussion of secularity and meaning systems, see Zuckerman et al. 2016. *The Nonreligious: Understanding Secular People and Societies*. New York: Oxford University Press.

9. Zuckerman, Phil. 2015. *Living the Secular life: New Answers to Old Questions*. London: Penguin Books.

10. Baker, Joseph O., and Buster G. Smith. 2015. *American Secularism: Cultural Contours of Nonreligious Belief Systems*. New York: New York University Press.

11. Shusterman, Richard. 2012. *Thinking through the Body: Essays in Somaesthetics*. Cambridge: Cambridge University Press, pg. 76.

12. Edgell, Penny, Joseph Gerteis, and Douglas Hartman. 2006. "Atheists as "Other": Moral Boundaries and Cultural Membership in American Society." *American Sociological Review* 71(2):211–234.

13. Gervais, Will M., Dimitris Xygalatas, Ryan T. McKay, Michiel van Elk, Emma E. Buchtel, Mark Aveyard, Sarah R. Schiavone, Ilan Dar-Nimrod, Annika M. Svedholm-Häkkinen, Tapani Riekki, Eva Kundtová Klocová, Jonathan E. Ramsay, and Joseph Bulbulia. 2017. "Global Evidence of Extreme Intuitive Moral Prejudice against Atheists." *Nature Human Behavior* 1(151):1–5.

14. Gray, John. 2018. *Seven Types of Atheism*. New York: Penguin Books, pg. 105.

15. RASIC_UK114, biology, professor, male, conducted March 4, 2011.

16. RASIC_US02, biology, graduate student, male, conducted March 2, 2015.

17. RASIC_UK13, biology, reader, female, conducted December 3, 2013.

18. RASIC_UK22, biology, lecturer, male, conducted December 4, 2013.

19. RASIC_US41, biology, professor, female, conducted April 3, 2015.

20. RASIC_UK07, biology, professor, male, conducted December 2, 2013.

21. RASIC_UK26, biology, lecturer, female, conducted December 4, 2013.

22. Baker, Joseph, and Buster G. Smith. 2015. *American Secularism: Cultural Contours of Nonreligious Belief Systems*. New York: New York University Press.

23. King, Michael, Louise Marsten, Sally McManus, Terry Brugha, Howard Meltzer, and Paul Bebbington. 2013. "Religion, Spirituality and Mental Health: Results from a National Study of English Households." *The Journal of British Psychiatry* 202(1):68–73.

24. Zuckerman, Phil. 2008. *Society Without God: What the Least Religious Nations Can Tell Us About Contentment*. New York: New York University Press.

25. RASIC_UK131, physics, research associate, male, conducted July 12, 2012.

26. RASIC_UK106, biology, lecturer, male, conducted March 1, 2011.

27. Hume, David. [1739–40]. 1978. *A Treatise of Human Nature*. Oxford: Oxford University Press.

28. RASIC_UK113, biology, emeritus senior research fellow, male, conducted March 3, 2011.

29. RASIC_US27, physics, graduate student, female, conducted April 1, 2015.

30. RASIC_UK108, biology, postdoc, male, conducted March 1, 2011.

31. RASIC_UK14, physics, professor, male, conducted December 3, 2013.

32. RASIC_US11, physics, professor, female, conducted March 25, 2015.

33. RASIC_UK20, physics, reader, male, conducted December 3, 2013.

34. RASIC_US19, physics, graduate student, male, conducted March 25, 2015.
35. RASIC_UK41, biology, lecturer, male, conducted December 6, 2013.
36. RASIC_UK23, physics, lecturer, male, conducted December 4, 2014.
37. RASIC_US30, biology, associate professor, female, conducted April 2, 2015.
38. RASIC_US60, biology, associate professor, female, conducted April 15, 2015.
39. RASIC_US25, biology, professor, male, conducted April 1, 2015.
40. RASIC_UK10, physics, postdoctoral fellow, female, conducted December 2, 2013.
41. Ruse, Michael. 2019. *A Meaning to Life*. New York: Oxford University Press, pg. 106.
42. Evans, John. 2018. *Morals Not Knowledge: Recasting the Contemporary U.S. Conflict Between Religion and Science*. Oakland: University of California Press.
43. RASIC_UK22, biology, lecturer, male, conducted December 4, 2013.
44. RASIC_US38, biology, associate professor, male, conducted April 3, 2015.
45. RASIC_UK34, biology, professor, male, conducted December 5, 2013.
46. RASIC_US22, physics, associate research scientist, male, conducted March 21, 2015.
47. RASIC_UK03, biology, principal investigator, female, conducted December 3, 2013.
48. RASIC_US60, biology, professor, female, conducted April 15, 2015.
49. RASIC_UK08, biology, professor, male, conduced December 2, 2013.
50. Gervais, Will M. 2014. "Everything Is Permitted? People Intuitively Judge Immorality as Representative of Atheists." PLoS ONE 9(4):e92302. Accessed November 20, 2019 (https://doi.org/10.1371/journal.pone.0092302).
51. Mudd, Tommy L., Maxine B. Najile, Ben K. L. Ng, and Will Gervais. 2015. "The Roots of Right and Wrong: Do Concepts of Innate Morality Reduce Intuitive Associations of Immorality With Atheism?" *Secularism and Nonreligion* 4(1):10.
52. RASIC_US01, physics, graduate student, male, conducted March 2, 2015.
53. RASIC_US26, physics, graduate student, female, conducted April 1, 2015.
54. Here we draw on the Theory of Basic Values by Schwartz, which has been confirmed in many different countries and cultures. Schwartz, Shalom H. 1992. "Universals in the Content and Structure of Values: Theoretical Advances and Empirical Tests in 20 Countries." *Advances in Experimental Social Psychology* 25:1–65; Schwartz, Shalom H., Gila Melech, Arielle Lehmann, Steven Burgess, Maris Harris, and Vicki Owens. 2001.

"Extending the Cross-Cultural Validity of the Theory of Basic Human Values with a Different Method of Measurement." *Journal of Cross-Cultural Psychology* 32(5):519–542.

55. RASIC_US02, physics, graduate student, male, conducted March 2, 2015.

56. For a nice summary of such research, see Zuckerman, Phil, Luke W. Galen, and Frank L. Pasquale. 2016. *The Nonreligious: Understanding Secular People and Societies.* New York: Oxford University Press.

57. RASIC_US03, biology, graduate student, female, conducted March 2, 2015.

58. RASIC_US26, physics, graduate student, female, conducted April 1, 2015.

59. Smith, Gregory A. 2017. "A Growing Share of Americans Say It's Not Necessary to Believe in God to Be Moral." *Pew Research Center,* October 16. Accessed July 17, 2020 (https://www.pewresearch.org/fact-tank/2017/10/16/a-growing-share-of-americans-say-its-not-necessary-to-believe-in-god-to-be-moral/).

60. Davies, Jim. 2018. "Religion Does Not Determine Your Morality." *The Conversation,* July 24. Accessed July 17, 2020 (https://theconversation.com/religion-does-not-determine-your-morality-97895).

61. RASIC_US60, biology, female, associate professor, conducted April 15, 2015.

Chapter 8

1. Barnes, Elizabeth M., Jasmine M. Truong, Daniel Z. Grunspan, and Sarah E. Brownell. 2020. "Are Scientists Biased Against Christians? Exploring Real and Perceived Bias Against Christians in Academic Biology." *Plos One.* Accessed July 24, 2020 (https://doi.org/10.1371/journal.pone.0226826).

2. Krause, Nicole M., Dominique Brossard, Dietram A. Scheufele, Michael A. Xenos, and Keith Franke. 2019. "Trends—Americans' Trust in Science and Scientists." *Public Opinion Quarterly* 83(4):817–836.

3. Ecklund, Elaine Howard, and Christopher P. Scheitle. 2018. *Religion Vs. Science: What Religious People Really Think.* New York: Oxford University Press.

4. Baker, Joseph O. 2012. "Public Perceptions of Incompatibility Between 'Science and Religion." *Public Understanding of Science* 21(3):340–353.

5. Emilsen, William W. 2012. "The New Atheism and Islam." *The Expository Times* 123(11):521–528.

6. Sturgis, Patrick, and Nick Allum. 2004. "Science in Society: Re-evaluating the Deficit Model of Public Attitudes." *Public Understanding of Science*

13(1):55–74; Bauer, Martin W., Nick Allum, and Steven Miller. 2007. "What Can We Learn from 25 Years of PUS Survey Research? Liberating and Expanding the Agenda." *Public Understanding of Science* 16(1):79–95.

7. Evans, John H. 2011. "Epistemological and Moral Conflict Between Religion and Science." *Journal for the Scientific Study of Religion* 50(4):707–727; Ecklund, Elaine Howard, and Christopher P. Scheitle. 2018. *Religion Vs. Science: What Religious People Really Think.* New York: Oxford University Press.

8. Leshner, Alan I. 2006. "Science and Public Engagement." *The Chronicle of Higher Education,* October 13. Accessed July 21, 2020(https://www.chron-icle.com/article/SciencePublic-Engagement/25084).

9. Scheufele, Dietram A. 2018. "Beyond the Choir? The Need to Understand Multiple Publics for Science." *Environmental Communication* 12(8): 1123–1126.

10. Evans, John H. 2011. "Epistemological and Moral Conflict Between Religion and Science." *Journal for the Scientific Study of Religion* 50(4):707–727; Johnson, David R., Christopher P. Scheitle, and Elaine Howard Ecklund. 2015. "Individual Religiosity and Orientation Towards Science: Reformulating Relationships." *Sociological Science* (2):106–124.

11. Blalock, Hubert M. 1967. *Toward a Theory of Minority Group Relations.* Hoboken, NJ: Wiley.

12. See the American Association for the Advancement of Science, Dialogue on Science, Ethics, and Religion here: https://www.aaas.org/programs/dialogue-science-ethics-and-religion.

13. BioLogos. "About Us." Accessed July 22, 2020(https://biologos.org/about-us).

14. BioLogos. 2018. "Expanding and Enriching: BioLogos Annual Report."

15. *Science for the Church.* "About Science for the Church." Accessed July 22, 2020(https://scienceforthechurch.org/about/). See also the website for Sinai and Synapses https://sinaiandsynapses.org/.

16. The Royal Society. 2019. In "Diversity Strategy 2019–2022," pg. 1. Accessed July 14, 2020(https://royalsociety.org/-/media/policy/topics/diversity-in-science/2019-09-Diversity-strategy-2019-22.pdf).

17. Bolger, Daniel, and Elaine Howard Ecklund. 2020. "Seeing Is Achieving: Religion, Embodiment, and Explanations of Racial Inequality in STEM." *Ethnic and Racial Studies* 59(2):269–288.

18. Huertas, Aaron. 2017. "What Does the Scientific Community Advocate For? The Case for 'Science Justice'." Medium, December 11 (https://medium.com/@aaronhuertas/what-does-the-scientific-community-advocate-for-the-case-for-science-justice-f6213f040cda).

Appendix

1. In addition to the book, findings from the RASIC study can be found in several articles, including: Ecklund, Elaine Howard, David R. Johnson, Christopher P. Scheitle, Kirstin R.W. Matthews, and Steven W. Lewis. 2016. "Religion among Scientists in International Context: A New Study of Scientists in Eight Regions." *Socius: Sociological Research for a Dynamic World* 2:1–9; Di, Di, and Elaine Howard Ecklund. 2017. "Producing Sacredness and Defending Secularity: Faith in the Workplace of Taiwanese Scientists." *Socius: Sociological Research for a Dynamic World* 3:1–15; Ecklund, Elaine Howard, David R. Johnson, Brandon Vaidyanathan, Kirstin R.W. Matthews, Steven W. Lewis, Robert A. Thomson Jr., and Di Di. 2019. *Secularity and Science: What Scientists around the World Really Think about Religion.* New York: Oxford University Press; Ecklund, Elaine Howard, and Di Di. 2018. "Global Spirituality among Scientists." In *Being Spiritual but Not Religious: Past, Present, Future(s)*, edited by W. B. Parsons. London and New York: Routledge; Ecklund, Elaine Howard, Christopher P. Scheitle, and Jared Peifer. 2018. "The Religiosity of Academic Scientists in the United Kingdom: Assessing the Role of Discipline and Department Status." *Journal for the Scientific Study of Religion* 57(4):743–757; Sorrell, Katherine, and Elaine Howard Ecklund. 2019. "How UK Scientists Legitimize Religion and Science Through Boundary Work." *Sociology of Religion* 80(3):350–371.
2. RASIC_UK108, biology, postdoctoral fellow, male, conducted March 1, 2011.
3. See Ecklund, Elaine Howard. 2010. *Science vs. Religion: What Scientists Really Think.* New York: Oxford University Press.
4. RASIC_UK114, biology, professor, male, conducted March 4, 2011.
5. RASIC_UK22, biology, lecturer, male, conducted December 4, 2013.
6. RASIC_UK22, biology, lecturer, male, conducted December 4, 2013.
7. RASIC_UK26, biology, lecturer, female, conducted December 4, 2013.
8. RASIC_US28, biology, postdoctoral fellow, male, conducted December 4, 2013
9. Hermanowicz, Joseph C. 2009. *Lives in Science: How Institutions Affect Academic Careers.* Chicago: University of Chicago Press.
10. Text entry not recorded for the U.K. survey.

References

Ammerman, Nancy T. 2013. *Sacred Stories, Spiritual Tribes: Finding Religion in Everyday Life*. New York: Oxford University Press.

Ammerman, Nancy T. 2013. "Spiritual But Not Religious? Beyond Binary Choices in the Study of Religion." *Journal for the Scientific Study of Religion* 52(2):258–278.

Atkins, Peter. 1981. *The Creation*. New York: W.H. Freeman & Co.

Aupers, Stef, and Dick Houtman. 2006. "Beyond the Spiritual Supermarket: The Social and Public Significance of New Age Spirituality." *Journal of Contemporary Religion* 21(2):201–222.

Baker, Joseph O. 2012. "Public Perceptions of Incompatibility Between 'Science and Religion.'" *Public Understanding of Science* 21(3):340–353.

Baker, Joseph O., and Buster G. Smith. 2015. *American Secularism: Cultural Contours of Nonreligious Belief Systems*. New York: Oxford University Press.

Bartkowski, John. 2004. *The Promise Keepers: Servants, Soldiers, and Godly Men*. New Brunswick, NJ: Rutgers University Press.

Barnes, Elizabeth M., Jasmine M. Truong, Daniel Z. Grunspan, and Sarah E. Brownell. 2020. "Are Scientists Biased Against Christians? Exploring Real and Perceived Bias Against Christians in Academic Biology." *Plos One*. Accessed July 24, 2020 (https://doi.org/10.1371/journal.pone.0226826).

Bauer, Martin W., Nick Allum, and Steven Miller. 2007. "What Can we Learn from 25 Years of PUS Survey Research? Liberating and Expanding the Agenda." *Public Understanding of Science* 16(1):79–95.

Beaman, Lori, and Steven Tomlins (Eds). 2015. *Atheist Identities: Spaces and Social Contexts*. New York: Springer International Publishing.

Bellah, Robert, Richard Madsen, William M. Sullivan, Ann Swidler, and Steven M. Tipton. 1985. *Habits of the Heart: Individualism and Commitment in American Life*. New York: Harper & Row.

Ben-David, Joseph, and Teresa A. Sullivan. 1975. "Sociology of Science." *Annual Review of Sociology* 1(1):203–222.

Berger, Peter. 2014. *The Many Altars of Modernity: Toward a Paradigm for Religion in a Pluralist Age*. Berlin: Walter de Gruyter GmbH & Co KG.

BioLogos. "About Us." Accessed July 22, 2020 (https://biologos.org/about-us).

BioLogos. 2018. "Expanding and Enriching: BioLogos Annual Report."

Blalock, Hubert M. 1967. *Toward a Theory of Minority Group Relations*. Hoboken, NJ: Wiley.

Blessing, Kimberly. 2013. "Atheism and the Meaningfulness of Life." In *The Oxford Handbook of Atheism*, edited by S. Bullivant and M. Ruse. New York: Oxford University Press, pp. 104–118.

Bolger, Daniel, and Elaine Howard Ecklund. 2020. "Seeing Is Achieving: Religion, Embodiment, and Explanations of Racial Inequality in STEM." *Ethnic and Racial Studies* (DOI: 10.1080/01419870.2020.1791354).

Bollinger, Alex. 2020. "Pastor Who Laid Hands on Trump Says Avoiding Coronavirus Is for 'Pansies.'" *LBGTQNation*, March 18. https://www.lgbtqnation.com/2020/03/pastor-laid-hands-trump-says-avoiding-coronavirus-pansies/.

Boudry, Maarten, and Massimo Pigliucci (Eds). 2017. *Science Unlimited? The Challenges of Scientism*. Chicago: University of Chicago Press.

Bourdieu, Pierre. 1973. "Cultural Reproduction and Social Reproduction." In *Knowledge, Education and Cultural Changes*, edited by R. Brown. London: Tavistock, pp. 71–112.

Bourdieu, Pierre. 1977. "Cultural Reproduction and Social Reproduction." In *Power and Ideology in Education*, edited by J. Karabel and A.H. Halsey. New York: Oxford University Press, pp. 487–511.

Bourdieu, Pierre. 1986. "The Forms of Capital." In *Handbook of Theory and Research for the Sociology of Education*, edited by J. Richardson. New York: Greenwood, pp. 46–58.

Browne, Janet. 2003. "Charles Darwin as a Celebrity." *Science in Context* 16: 175–194.

Bruce, Steve. 2002. *God Is Dead: Secularization in the West*. Malden, MA: Wiley-Blackwell.

Burkeman, Oliver. 2013. "'Scientism' Wars: There's an Elephant in the Room, and Its Name Is Sam Harris." August 27. *The Guardian*. Accessed July 24, 2020 (https://www.theguardian.com/news/oliver-burkeman-s-blog/2013/aug/27/scientism-wars-sam-harris-elephant).

Chesterton, G.K. 1910. *What's Wrong with the World*. Mineola, NY: Dover Publications.

Cohen, Steven M., and Arnold M. Eisen. 2000. *The Jew Within: Self, Family, and Community in America*. Bloomington: Indiana University Press.

Collins, Francis. 2006. *The Language of God: A Scientist Presents Evidence for Belief*. New York: Free Press.

Collins, Randall. 1993. "Emotional Energy as the Common Denominator of Rational Action." *Rationality and Society* 5(2):203–230.

Collins, Randall. 1998. *The Sociology of Philosophies: A Global Theory of Intellectual Change*. Cambridge, MA: Belknap Press of Harvard University Press.

Connor, Steve. 2011. "For the Love of God . . . Scientists in Uproar at £1m Religion Prize." *The Independent*, April 7. Accessed January 15, 2015 (http://www.independent.co.uk/news/science/for-the-love-of-god-scientists-in-uproar-at-1631m-religion-prize-2264181.html).

Cornwall Alliance. 2010. "Sounding the Alarm about Dangerous Environmental Extremism: Explosive New DVD Series, Resisting the Green Dragon, Now Being Distributed Nationally and Abroad." *Cornwall Alliance.* Accessed June 23, 2020 (https://cornwallalliance.org/2010/11/sounding-the-alarm-about-dangerous-environmental-extremism-explosive-new-dvd-series-resisting-the-green-dragon-now-being-distributed-nationally-and-abroad/).

Cottee, Simon. 2015. *The Apostates: When Muslims Leave Islam.* London: Hurst & Company.

Cragun, Ryan T., Barry Kosmin, Ariela Keysar, Joseph H. Hammer, and Michael Nielsen. 2012. "On the Receiving End: Discrimination Toward the Non-religious in the United States." *Journal of Contemporary Religion* 27(1):105–127.

Crick, Francis. 1994. *The Astonishing Hypothesis: The Scientific Search for the Soul.* New York: Touchstone.

Daum, Andreas W. 2009. "Varieties of Popular Science and the Transformations of Public Knowledge: Some Historical Reflections." *Isis* 100:319–332.

Davie, Grace. 1990. "Believing without Belonging: Is This the Future of Religion in Britain?" *Social Compass* 37(4):455–469.

Davie, Grace. 2007. *The Sociology of Religion.* London: Sage.

Davies, Jim. 2018. "Religion Does Not Determine your Morality." *The Conversation,* July 24. Accessed July 17, 2020 (https://theconversation.com/religion-does-not-determine-your-morality-97895).

Dawkins, Richard. 1989. *The Selfish Gene.* Oxford: Oxford University Press.

Dawkins, Richard. 1992. "A Scientist's Case Against God." Presented at Edinburgh International Science Festival, April 15, Edinburgh, UK.

Dawkins, Richard. 2006. *The God Delusion.* London: Bantam Press.

Dawkins, Richard. 2011. "The Virus of Faith." Documentary UK Channel 4. Accessed October 21, 2019 (https://www.youtube.com/watch?v=eVy-0E1x620).

Dawkins, Richard. 2015. "Don't Force Your Religious Opinions on Your Children." *Time,* February 19.

Dein, Simon. 2013. "The Origins of Jewish Guilt: Psychological, Theological, and Cultural Perspectives." *Journal of Spirituality in Mental Health* 15(2):123–137.

Demarath III, N. J. 2000. "The Rise of 'Cultural Religion' in European Christianity: Learning from Poland, Northern Ireland, and Sweden." *Social Compass* 47(1):127–139.

Dennett, Daniel C. 2006. "Common-Sense Religion." *The Chronicle of Higher Education* 52(20):B6.

De Ridder, Jeroen, Rik Peels, and Rene van Woudenberg (Eds). 2018. *Scientism: Prospects and Problems.* Oxford: Oxford University Press.

Di, Di, and Elaine Howard Ecklund. 2017. "Producing Sacredness and Defending Secularity: Faith in the Workplace of Taiwanese Scientists." *Socius: Sociological Research for a Dynamic World* 3:1–15.

Di Fiore, James. 2016. "My Kids Will Be Raised Without Religion." *Huffpost*, April 25.

Dillon, Michele, and Paul Wink. 2007. *In the Course of a Lifetime: Tracing Religious Belief, Practice, and Change.* Berkeley: University of California Press.

Doane, Michael J., and Marta Elliot. 2015. "Perceptions of Discrimination Among Atheists: Consequences for Atheist Identification, Psychological and Physical Well-Being." *Psychology of Religion and Spirituality* 7(2):130–142.

Draper, Scott, and Joseph O. Baker. 2011. "Angelic Belief as American Folk Religion." *Sociological Forum* 26(3):623–43.

Du Bois, W. E. B. 1940. *Dusk of Dawn: An Essay Toward an Autobiography of a Race Concept.* New York: Harcourt Brace.

Eaton, Marc A. 2015. "'Give Us a Sign of Your Presence': Paranormal Investigation as a Spiritual Practice." *Sociology of Religion* 76(4):389–412.

Ecklund, Elaine Howard. 2010. *Science vs. Religion: What Scientists Really Think.* New York: Oxford University Press.

Ecklund, Elaine Howard, and Christopher P. Scheitle. 2018. *Religion vs. Science: What Religious People Really Think.* New York: Oxford University Press.

Ecklund, Elaine Howard, Christopher P. Scheitle, and Jared Peifer. 2018. "The Religiosity of Academic Scientists in the United Kingdom: Assessing the Role of Discipline and Department Status." *Journal for the Scientific Study of Religion* 57(4):743–757.

Ecklund, Elaine Howard, and Di Di. 2018. "Global Spirituality Among Scientists." In *Being Spiritual but Not Religious: Past, Present, Future(s)*, edited by W. B. Parsons. London and New York: Routledge.

Ecklund, Elaine Howard, David R. Johnson, Christopher P. Scheitle, Kirstin R.W. Matthews, and Steven W. Lewis. 2016. "Religion among Scientists in International Context: A New Study of Scientists in Eight Regions." *Socius: Sociological Research for a Dynamic World* 2:1–9.

Ecklund, Elaine Howard, David R. Johnson, Brandon Vaidyanathan, Kirstin R.W. Matthews, Steven W. Lewis, Robert A. Thomson Jr., and Di Di. 2019. *Secularity and Science: What Scientists around the World Really Think about Religion.* New York: Oxford University Press.

Ecklund, Elaine Howard, and Elizabeth Long. 2011. "Scientists and Spirituality." *Sociology of Religion* 72:253–274.

Ecklund, Elaine Howard, and Kristen Schultz Lee. 2011. "Atheists and Agnostics Negotiate Religion and Family." *Journal for the Scientific Study of Religion* 50(4):728–743.

Egdell, Penny, Joseph Gerteis, and Douglas Hartmann. 2006. "Atheists as 'Other': Moral Boundaries and Cultural Membership in American Society." *American Sociological Review* 72(2):211–234.

Emilsen, William W. 2012. "The New Atheism and Islam." *The Expository Times* 123(11):521–528.

Encyclopedia Britannica. 2018. "United Kingdom." Accessed April 27, 2018 (http://www.britannica.com/EBchecked/topic/615557/United-Kingdom/44685/Religion).

Evans, John H. 2011. "Epistemological and Moral Conflict between Religion and Science." *Journal for the Scientific Study of Religion* (50)4:707–727.

Evans, John. 2018. *Morals Not Knowledge: Recasting the Contemporary U.S. Conflict Between Religion and Science.* Oakland: University of California Press.

Fahy, Declan. 2015. *The New Celebrity Scientists: Out of the Lab and into the Limelight.* Lanham, MD: Rowman & Littlefield.

Focus on the Family. 2011. "Marriage Between an Atheist and a Christian." (https://www.focusonthefamily.com/family-qa/marriage-between-an-atheist-and-a-christian/).

Frost, Jacqui. 2017. "Rejecting Rejection Identities: Negotiating Positive Non-religiosity at the Sunday Assembly." In *Religion and Its Others: Studies in Religion, Nonreligion, and Secularity, Volume 6*, edited by S. Gutkowski, L. Lee, and J. Quack. Boston and Berlin: Walter de Gruyter.

Frost, Jacqui. 2019. "Certainty, Uncertainty, or Indifference? Examining Variation in the Identity Narratives of Nonreligious Americans." *American Sociological Review* 84(5):828–850.

Frost, Jaqui. 2020. "Review of *A Qualitative Study of Black Atheists: 'Don't Tell Me You're One of Those!'*" *Social Forces* 99(2):e1–e3.

Garelli, Franco. 2014. *Religion Italian Style: Continuities and Change in a Catholic Community.* Abingdon-on-Thames, UK: Routledge.

Garelli, Franco. 2016. "Preface." In *Annual Review of the Sociology of Religion: Sociology of Atheism*, edited by R. Cipriani and F. Garelli. Leiden: Brill.

Gervais, Will M. 2014. "Everything Is Permitted? People Intuitively Judge Immorality as Representative of Atheists." PLoS ONE 9(4): e92302. Accessed November 20, 2019 (https://doi.org/10.1371/journal.pone.0092302).

Gervais, Will M., Dimitris Xygalatas, Ryan T. McKay, Michiel van Elk, Emma E. Buchtel, Mark Aveyard, Sarah R. Schiavone, Ilan Dar-Nimrod, Annika M. Svedholm-Häkkinen, Tapani Riekki, Eva Kundtová Klocová, Jonathan E. Ramsay, and Joseph Bulbulia. 2017. "Global Evidence of Extreme Intuitive Moral Prejudice against Atheists." *Nature Human Behavior* 1(151):1–5.

Gervais, Will M., and Maxine Najile. 2018. "How Many Atheists Are There?" *Social Psychological and Personality Science* 9(1):3–10.

Giberson, Karl, and Mariano Artigas. 2007. *Oracles of Science: Celebrity Scientists versus God and Religion.* New York: Oxford University Press.

Gieryn, Thomas F. 1999. *Cultural Boundaries of Science: Credibility on the Line.* Chicago: University of Chicago Press.

Glass, Jennifer L., April Sutton, and Scott T. Fitzgerald. 2015. "Leaving the Faith: How Religious Switching Changes Pathways to Adulthood among Conservative Protestant Youth." *Social Currents* 2(2):126–143.

Gleiser, Marcelo. 2014. *The Island of Knowledge: The Limits of Science and the Search for Meaning.* New York: Basic Books.

Goffman, Erving. 1963. *Behavior in Public Places.* Glencoe, IL: Free Press of Glencoe.

Goldberg, Jonah. 2020. "The Treason of Epidemiologists." *The Dispatch.* Accessed June 23, 2020 (https://gfile.thedispatch.com/p/the-treason-of-epidemiologists).

Goldstein, Eric L. 1997. "'Different Blood Flows in Our Veins': Race and Jewish Self-Definition in Late-Nineteenth-Century America." *American Jewish History* 85:29–55.

Gould, Stephen Jay. 1997. "Nonoverlapping Magisteria." *Natural History* 106:16–22.

Gray, John. 2018. *Seven Types of Atheism.* New York: Penguin Books.

Haberman, Clyde. 2014. "From Private Ordeal to National Fight: The Case of Terri Schiavo." *New York Times*, April 20. Accessed July 1, 2020 (https://www.nytimes.com/2014/04/21/us/from-private-ordeal-to-national-fight-the-case-of-terri schiavo.html#:~:text=For%2015%20years%2C%20Terri%20Schiavo,turned%20her%20ordeal%20into%20a).

Habermas, Jurgen. 1971. *Knowledge and Human Interests.* Cambridge, MA: Polity Press.

Hamburger, Philip. 2009. *Separation of Church and State.* Cambridge, MA: Harvard University Press.

Hammer, Joseph H., Ryan T. Cragun, Karen Hwang, and Jesse M. Smith. 2012. "Forms, Frequency, and Correlates of Perceived Anti-atheist Discrimination." *Secularism and Nonreligion* 1:43–67.

Harris, Sam. 2006. "Day 1 (Sam Harris): Why Are Atheists So Angry?" *Jewcy*, November 16. (https://jewcy.com/jewish-religion-and-beliefs/monday_why_are_atheists_so_angry_sam_harris).

Harris, Sam. 2006. "Science Must Destroy Religion." *Edge*, January 2. (https://www.edge.org/response-detail/11122).

Harris, Sam. 2008. *Letter to a Christian Nation.* New York: Vintage Books.

Harris, Sam. 2010. *The Moral Landscape: How Science Can Determine Human Values.* New York: Free Press.

Hawking, Stephen. 1996. *A Brief History of Time: The Updated and Expanded Tenth Anniversary Edition.* New York: Bantam Books.

Hawking, Stephen, and Leonard Mlodinow. 2010. *The Grand Design*. New York: Bantam Books.

Heelas, Paul. 2006. "Challenging Secularization Theory: The Growth of 'New Age' Spiritualties of Life." *Hedgehog Review* 8(1):46.

Heelas, Paul, and Linda Woodhead 2005. *The Spiritual Revolution: Why Religion Is Giving Way to Spirituality*, edited by B. Seel, B. Szerszynski, and K. Tusting. Malden, MA: Blackwell Publishing.

Hemming, Peter J., and Christopher Roberts. 2018. "Church Schools, Educational Markets and the Rural Idyll." *British Journal of Sociology of Education* 39(4):501–517.

Hermanowicz, Joseph C. 2009. *Lives in Science: How Institutions Affect Academic Careers*. Chicago: University of Chicago Press.

Hill, Jonathan P., and Brandon Vaidyanathan. 2011. "Substitution or Symbiosis? Assessing the Relationship Between Religious and Secular Giving." *Social Forces* 90(1):157–180.

Hitchens, Christopher, Richard Dawkins, Sam Harris, and Daniel Dennett (forward by Stephen Fry). 2019. *The Four Horsemen: The Conversation that Sparked an Atheist Revolution*. New York: Random House.

Horton, Richard. 2000. "Genetically Modified Food: Consternation, Confusion, and Crack-Up." *Medical Journal of Australia* 177:148–149.

Huertas, Aaron. 2017. "What Does the Scientific Community Advocate For? The Case for 'Science Justice.'" Medium, December 11. (https://medium.com/@aaronhuertas/what-does-the-scientific-community-advocate-for-the-case-for-science-justice-f6213f040cda).

Hume, David. [1739–40]. 1978. *A Treatise of Human Nature*. Oxford: Oxford University Press.

Ipsos. 2010. "Ipsos Global @dvisory: Is Religion a Force for Good in the World? Combined Population of 23 Major Nations Evenly Divided in Advance of Blair, Hitchens Debate." November 25. Accessed July 15, 2020 (https://www.ipsos.com/en-us/news-polls/ipsos-global-dvisory-religion-force-good-world-combined-population-23-major-nations-evenly-divided).

Jeffries, Stuart. 2007. "Britain's New Cultural Divide Is Not Between Christian and Muslim, Hindu, and Jew. It Is Between Those Who Have Faith and Those Who Don't." *The Guardian*, February 26.

Johnson, David R. 2017. *A Fractured Profession: Commercialism and Conflict in Academic Science*. Baltimore, MD: Johns Hopkins University Press.

Johnson, David R., Christopher P. Scheitle, and Elaine Howard Ecklund. 2015."Individual Religiosity and Orientation Towards Science: Reformulating Relationships." *Sociological Science* 2:106–124.

Johnson, David R., Elaine Howard Ecklund, Di, and Kirstin R.W. Matthews. 2018. "Responding to Richard: Celebrity and (Mis)representation of Science." *Public Understanding of Science* 27(5):535–549.

Johnson, David R., and Jared L. Peifer. 2017. "How Public Confidence in Higher Education Varies by Social Context." *The Journal of Higher Education* 88(4):619–644.

Johnson, David R., and Joseph C. Hermanowicz. 2017. "Peer Review: From 'Sacred Ideals' to 'Profane Realities.'" In *Higher Education: Handbook of Theory and Research*. Switzerland: Springer, pp. 485–527.

Keller, Tim. 2009. "Tim Keller on the New Atheists." *Big Think*. Accessed June 25, 2020 (https://bigthink.com/videos/tim-keller-on-the-new-atheists).

Kettell, Steve. 2016. "What's Really New about New Atheism." *Palgrave Communications* 2:16099.

Kierkegaard, Soren. 1846. *Concluding Postscripts to Scientific Inquiry* in *Philosophical Fragments*. Tr. Howard V. Hong and Edna H. Hong, 1992. Princeton, NJ: Princeton University Press.

King, Michael, Louise Marsten, Sally McManus, Terry Brugha, Howard Meltzer, and Paul Bebbington. 2013. "Religion, Spirituality and Mental Health: Results from a National Study of English Households." *The Journal of British Psychiatry* 202(1):68–73.

Kosmin, Barry Alexander, and Ariela Keysar. 2009. *American Religious Identification Survey (ARIS 2008): Summary Report*. Trinity College.

Kraus, Rachel. 2009. "Straddling the Sacred and Secular: Creating a Spiritual Experience Through Belly Dance." *Sociological Spectrum* 29(5):598–625.

Krause, Nicole M., Dominique Brossard, Dietram A. Scheufele, Michael A. Xenos, and Keith Franke. 2019. "Trends—Americans' Trust in Science and Scientists." *Public Opinion Quarterly* 83(4):817–836.

Lamont, Miche`le. 2001. "Culture and Identity." In *Handbook of Sociological Theory*, edited by J. H. Turner. New York: Plenum, pp. 171–186.

Lareau, Annette. 2003. *Unequal Childhoods: Class, Race, and Family Life*. Berkeley: University of California Press.

Larson, Edward J., and Larry Witham. 1998. "Leading Scientists Still Reject God." *Nature* 394(313).

Lee, Lois. 2015. *Recognizing the Non-Religious*. Oxford: Oxford University Press.

Leshner, Alan I. 2006. "Science and Public Engagement." *The Chronicle of Higher Education*, October 13. Accessed July 21, 2020 (https://www.chronicle.com/article/SciencePublic-Engagement/25084).

Leuba, James H. 1934. "Religious Beliefs of American Scientists." *Harper's Magazine* 169:291–300.

Lipka, Michael. 2015. "A Closer Look at America's Rapidly Growing Religious 'Nones.'" *Pew Research Forum*. (https://www.pewresearch.org/fact-tank/2015/05/13/a-closer-look-at-americas-rapidly-growing-religious-nones/).

Lipka, Michael, and Claire Gecewicz. 2017. "More Americans Now Say They're Spiritual but Not Religious." *Pew Research Center*, September 6. Accessed July 13, 2020 (https://www.pewresearch.org/fact-tank/2017/09/06/more-americans-now-say-theyre-spiritual-but-not-religious/).

Marler, Penny Long, and C. Kirk Hadaway. 2002. "'Being Religious' or 'Being Spiritual' in America: A Zero-Sum Proposition?" *Journal for the Scientific Study of Religion* 41:289–300.

Marsden, Peter V. 1988. "Homogeneity in Confiding Relations." *Social Networks* 10:57–76.

Martin, Richard C. 2010. "Hidden Bodies in Islam: Secular Muslim Identities in Modern (and Premodern) Societies." In *Muslim Societies and the Challenge of Secularization: An Interdisciplinary Approach*, edited by G. Marranci. Switzerland: Springer, pp. 131–148.

Mayrl, Damon, and Freeden Oeur. 2009. "Religion and Higher Education: Current Knowledge and Directions for Future Research." *Journal for the Scientific Study of Religion* 48(2):260–275.

McMahan, David L. 2004. "Buddhism: Introducing the Buddhist Experience (Review)." *Philosophy East and West* 54(2):268–270.

McPherson, J. Miller, and Lynn Smith-Lovin. 1987. "Homophily in Voluntary Organizations: Status Distance and the Composition of Face to Face Groups." *American Sociological Review* 52:370–379.

McPherson, J. Miller, Lynn Smith-Lovin, and James M. Cook. 2001. "Birds of a Feather: Homophily in Social Networks." *Annual Review of Sociology* 27:415–444.

Mercandante, Linda A. 2014. *Belief Without Borders: Inside the Minds of the Spiritual but Not Religious*. New York: Oxford University Press.

Merton, Robert K. 1957. "Priorities in Scientific Discovery: A Chapter in the Sociology of Science." *American Sociological Review* 22(6):635–659.

Merton, Robert K. 1973. *The Sociology of Science: Theoretical and Empirical Investigations*. Chicago: University of Chicago Press.

Millard, Erickson J. 1983. *Christian Theology*. Grand Rapids, MI: Baker Academic.

Mitroff, Ian. 1974. "Norms and Counter-Norms in a Select Group of the Apollo Moon Scientists: A Case Study of the Ambivalence of Scientists." *American Sociological Review* 39(4):579–595.

Mudd, Tommy L., Maxine B. Najile, Ben K. L. Ng, and Will Gervais. 2015. "The Roots of Right and Wrong: Do Concepts of Innate Morality Reduce Intuitive Associations of Immorality with Atheism?" *Secularism and Nonreligion* 4(1):10.

Numbers, Ronald L., and Jeff Hardin. 2018. *The Warfare Between Science and Religion: The Idea That Wouldn't Die*, edited by J. Hardin, R. L. Numbers, and R. A. Blinzley. Baltimore, MD: Johns Hopkins University Press.

Nye, Bill. 2016. "Hey Bill Nye, 'Does Science Have All the Answers or Should We Do Philosophy Too?" *Big Think*. (https://www.youtube.com/watch?v=ROe28Ma_tYM&feature=youtu.be).

Park, Alison, Caroline Bryson, Elizabeth Clery, John Curtice, and Miranda Philips. 2013. *British Social Attitudes: The 30th Report*. London: NatCen Social Research.

Parsons, Talcott, and Gerald Platt. 1968. *The American Academic Profession: A Pilot Study*. National Science Foundation.

Peterson, Gregory R. 2003. "Demarcation and the Scientistic Fallacy." *Zygon Journal of Religion and Science* 38(4):751–761.

Pew Research Center. 2015. "America's Changing Religious Landscape." (https://www.pewforum.org/2015/05/12/americas-changing-religious-landscape/).

Pew Research Center. 2018. "Attitudes Toward Spirituality and Religion" in "Being Christian in Western Europe." *Pew Research Center*, May 29. Accessed July 15, 2020 (https://www.pewforum.org/2018/05/29/attitudes-toward-spirituality-and-religion/).

Pew Research Center. 2019. "Americans Have Positive Views about Religion's Role in Society, but Want It Out of Politics." *Pew Research Center*, November 15 (https://www.pewforum.org/2019/11/15/americans-have-positive-views-about-religions-role-in-society-but-want-it-out-of-politics/).

Pigliucci, Massimo. 2014. "Neil deGrasse Tyson and the Value of Philosophy." *Scientia Salon*, May 12. (https://scientiasalon.wordpress.com/2014/05/12/neil-degrasse-tyson-and-the-value-of-philosophy/).

Putnam, Robert D., and David E. Campbell. 2010. *American Grace: How Religion Divides and Unites Us*. New York: Simon & Schuster.

Reddit. 2017. "Atheist Married to Christian." (https://www.reddit.com/r/atheism/comments/56munj/atheist_married_to_christian/).

Rienzo, Cinzia, and Carlos Vargas-Silva. 2015. "Targeting Migration with Limited Control: The Case of the UK and the EU." *IZA Journal of European Labor Studies* 4(16).

Roof, Wade Clark. 1998. "Religious Borderlands: Challenges for Future Study." *Journal for the Scientific Study of Religion* 37:1–14.

Roof, Wade Clark. 1999. *Spiritual Marketplace: Baby Boomers and the Remaking of American Religion*. Princeton, NJ: Princeton University Press.

Rosenberg, Alex. 2011. *The Atheist's Guide to Reality: Enjoying Life Without Illusions*. New York: W.W. Norton & Company.

The Royal Society. 2019. "Diversity Strategy 2019–2022." Accessed July 14, 2020 (https://royalsociety.org/-/media/policy/topics/diversity-in-science/2019-09-Diversity-strategy-2019-22.pdf).

Rubenstein, Richard. 1966. *After Auschwitz: Radical Theology and Contemporary Judaism*. Indianapolis, IN: Bobbs-Merrill.

Ruse, Michael. 2019. *A Meaning to Life*. New York: Oxford University Press.

Sagan, Carl. 1980. *Cosmos*. New York: Ballantine Books.

Sahgal, Neha. 2018. "10 Key Findings about Religion in Western Europe." *Pew Research Center*, May 29. Accessed July 15, 2020 (https://www.pewresearch.org/fact-tank/2018/05/29/10-key-findings-about-religion-in-western-europe/).

Scheitle, Christopher P., and Elaine Howard Ecklund. 2017. "Recommending a Child Enter a STEM Career: The Role of Religion." *Journal of Career Development* 44(3):251–265.

Scheufele, Dietram A. 2018. "Beyond the Choir? The Need to Understand Multiple Publics for Science." *Environmental Communication* 12(8):1123–1126.

Schmidt, Leigh Eric. 2016. *Village Atheists: How American Unbelievers Made Their Way in A Godly Nation.* Princeton, NJ: Princeton University Press.

Schnable, Allison. 2015. "Religion and Giving for International Aid: Evidence from a Survey of US Church Members." *Sociology of Religion* 76(1):72–94.

Schulson, Michael. 2014. "Jonathan Sacks on Richard Dawkins: 'New Atheists Lack a Sense of Humor." *Salon.* Accessed June 25, 2020 (https://www.salon.com/2014/09/27/jonathan_sacks_on_richard_dawkins_new_atheists_lack_a_sense_of_humor/).

Schwartz, Shalom H. 1992. "Universals in the Content and Structure of Values: Theoretical Advances and Empirical Tests in 20 Countries." *Advances in Experimental Social Psychology* 25:1–65.

Schwartz, Shalom H., Gila Melech, Arielle Lehmann, Steven Burgess, Maris Harris, and Vicki Owens. 2001. "Extending the Cross-Cultural Validity of the Theory of Basic Human Values with a Different Method of Measurement." *Journal of Cross-Cultural Psychology* 32(5):519–542.

Science for the Church. "About Science for the Church." Accessed July 22, 2020 (https://scienceforthechurch.org/about/).

Sedghi, Ami. 2013. "UK Census: Religion by Age, Ethnicity and Country of Birth," *The Guardian,* May 16.

Shusterman, Richard. 2012. *Thinking through the Body: Essays in Somaesthetics.* Cambridge: Cambridge University Press.

Simon, Scott. 2017. "Richard Dawkins on Terrorism and Religion." *NPR,* May 27.

Slisco, Aila. 2020. "Pastor Holds Service with Over 1,000 Parishioners in Defiance of Large-Gathering Ban." *Newsweek,* March 18. (https://www.newsweek.com/pastor-holds-service-over-1000-parishoners-defiance-large-gathering-ban-1493113).

Smith, Jesse M. 2011. "Becoming an Atheist in America: Constructing Identity and Meaning from the Rejection of Theism." *Sociology of Religion* 72(2):215–237.

Smith, Jesse M. 2013. "Creating a Godless Community: The Collective Identity Work of Contemporary American Atheists." *Journal for the Scientific Study of Religion* 52(1):80–81.

Smith, Gregory A. 2017. "A Growing Share of Americans Say It's Not Necessary to Believe in God to Be Moral." *Pew Research Center,* October 16. Accessed July 17, 2020 at https://www.pewresearch.org/fact-tank/2017/10/16/a-growing-share-of-americans-say-its-not-necessary-to-believe-in-god-to-be-moral/.

Sorrell, Katherine, and Elaine Howard Ecklund. 2019. "How UK Scientists Legitimize Religion and Science Through Boundary Work." *Sociology of Religion* 80(3):350–371.

Stark, Rodney. 1963. "On the Incompatibility of Religion and Science: A Survey of American Graduate Students." *Journal for the Scientific Study of Religion* 3:3–20.

Stark, Rodney, and Roger Finke. 2000. *Acts of Faith: Explaining the Human Side of Religion*. Berkeley: University of California Press.

Stenger, Victor. 2007. God: *The Failed Hypothesis—How Science Shows that God Does Not Exist*. Amherst, MA: Prometheus Books.

Stenmark, Mikael. 2001. *Scientism: Science, Ethics, and Religion*. Aldershot, UK: Ashgate.

Stirrat, Michael, and R. Elizabeth Cornwell. 2013. "Eminent Scientists Reject the Supernatural: A Survey of the Fellows of the Royal Society." *Evolution: Education and Outreach* 6(1):33.

Sturgis, Patrick, and Nick Allum. 2004. "Science in Society: Re-evaluating the Deficit Model of Public Attitudes." *Public Understanding of Science* 13(1):55–74.

Swann, Daniel. 2020. *A Qualitative Study of Black Atheists: "Don't Tell Me You're One of Those!"* London: Rowman and Littlefield.

Tenenbaum, Shelly, and Lynn Davidman. 2007. "It's in My Genes: Biological Discourse and Essentialist Views of Identity among Contemporary American Jews." *The Sociological Quarterly* 48(3):435–450.

Underwood, Lynn G., and Jeanne A. Teresi. 2002. "The Daily Spiritual Experience Scale: Development, Theoretical Description, Reliability, Exploratory Factor Analysis, and Preliminary Validity Using Health-Related Data." *Annual Behavioral Medicine* 24:22–33.

Vaidyanathan, Brandon, Hill, Jonathan P., and Smith, Christian. 2011. "Religion and Charitable Financial Giving to Religious and Secular Causes: Does Political Ideology Matter?" *Journal for the Scientific Study of Religion* 50(3):450–469.

Voas, David, and Alasdair Crockett. 2005. "Religion in Britain: Neither Believing nor Belonging." *Sociology* 39(1):11–28.

Watson, P. J., and Ronald Morris. 2005. "Spiritual Experience and Identity: Relationships with Religious Orientation, Religious Interest, and Intolerance of Ambiguity." *Review of Religious Research* 46:371–79.

Weinberg, Steven. 1999. "A Designer Universe?" *Conference on Cosmic Design of the American Association for the Advancement of Science*. Accessed June 29, 2020 (https://www.physlink.com/Education/essay_weinberg.cfm).

Wilson, Edward O. 1978. *On Human Nature*. Cambridge, MA: Harvard University Press.

Wood, Jack, and Gianpiero Petriglieri. 2005. "Transcending Polarization: Beyond Binary Thinking." *Transactional Analysis Journal* 35(1): 31–39.

Wright, N.T. 2013. *Paul and the Faithfulness of God*. Minneapolis, MN: Fortress Press.

Wuthnow, Robert. 1989. *Communities of Discourse: Ideology and Social Structure in the Reformation, the Enlightenment, and European Socialism*. Cambridge, MA: Harvard University Press.

Wuthnow, Robert. 1998. *After Heaven: Spirituality in America since the 1950s*. Berkeley and Los Angeles: University of California Press.

Wuthnow, Robert. 2002. *Loose Connections*. Cambridge, MA: Harvard University Press.

Wuthnow, Robert. 2020. *What Happens When We Practice Religion? Textures of Devotion in Everyday Life*. Princeton, NJ: Princeton University Press.

Wuthnow, Robert, and Wendy Cadge. 2004. "Buddhists and Buddhism in the United States: The Scope of Influence." *Journal for the Scientific Study of Religion* 43:363–80.

Zuckerman, Phil. 2008. *Society Without God: What the Least Religious Nations Can Tell Us About Contentment*. New York: New York University Press.

Zuckerman, Phil. 2012. *Faith No More: Why People Reject Religion*. New York: Oxford University Press.

Zuckerman, Phil. 2015. *Living the Secular life: New Answers to Old Questions*. London: Penguin Books.

Zuckerman, Phil, Luke W. Galen, and Frank L. Pasquale. 2016. *The Nonreligious: Understanding Secular People and Societies*. New York: Oxford University Press.

Index

For the benefit of digital users, indexed terms that span two pages (e.g., 52–53) may, on occasion, appear on only one of those pages.

Tables are indicated by *t* following the page number